AF801642

Manhattan Review®

Test Prep & Admissions Consulting

Turbocharge Your DSAT: Math Essentials

Published on February 11th, 2025

Version 70

- *Digital SAT specific concepts and theory*
- *Intuitive and graphical explanations of concepts*
- *Smart math Shortcuts*
- *Comprehensive collection of math formulae*
- *Over 100 practice questions*
- *Mapped according to the scope of the Digital SAT*

www.manhattanreview.com

Copyright and Terms of Use

Copyright and Trademark

Terms of Use

No Warranties

Limitation on Liability

International Standard Book Number (ISBN): 978-1-62926-115-7

Published on February 11th, 2025.

Manhattan Review, 2 Park Avenue, Suite 2010, New York, NY 10016.
Phone: +1 (212) 316-2000. E-Mail: info@manhattanreview.com. Web: www.manhattanreview.com

About the Turbocharge your SAT Series

The Turbocharge Your SAT Series was created to provide students with comprehensive and highly effective SAT preparation for maximum SAT performance. Thousands of students around the world have received substantial score improvements by using Manhattan Review's SAT prep books. Now in its updated 2nd edition for the new SAT in 2016, the full series of 12 guides is designed to provide SAT students with rigorous, thorough, and accessible SAT instruction for top SAT scores. Manhattan Review's SAT prep books precisely target each testing area and deconstruct the different test sections in a manner that is both student-centered and results-driven, teaching test-takers everything they need to know in order to significantly boost their scores. Covering all of the necessary material in mathematics and verbal skills from the most basic through the most advanced levels, the Turbocharge Your SAT Series is the top study resource for all stages of SAT preparation. Students who work through the complete series develop all of the skills, knowledge, and strategies needed for their best possible SAT scores.

- ■ **DSAT Math Essentials (ISBN: 978-1-62926-115-7)**
- □ **DSAT Math Guide (ISBN: 978-1-62926-116-4)**
- □ **DSAT Reading & Writing Guide (ISBN: 978-1-62926-117-1)**
- □ **DSAT Math Question Bank (ISBN: 978-1-62926-118-8)**

About the Company

Manhattan Review's origin can be traced directly to an Ivy-League MBA classroom in 1999. While lecturing on advanced quantitative subjects to MBAs at Columbia Business School in New York City, Prof. Dr. Joern Meissner was asked by his students to assist their friends, who were frustrated with conventional GMAT preparation options. He started to create original lectures that focused on presenting the GMAT content in a coherent and concise manner rather than a download of voluminous basic knowledge interspersed with so-called "tricks." The new approach immediately proved highly popular with GMAT students, inspiring the birth of Manhattan Review. Over the past 15+ years, Manhattan Review has grown into a multi-national firm, focusing on GMAT, GRE, LSAT, SAT, and TOEFL test prep and tutoring, along with business school, graduate school and college admissions consulting, application advisory and essay editing services.

About the Founder

Professor Joern Meissner, the founder and chairman of Manhattan Review has over twenty-five years of teaching experience in undergraduate and graduate programs at prestigious business schools in the USA, UK and Germany. He created the original lectures, which are constantly updated by the Manhattan Review Team to reflect the evolving nature of the GMAT GRE, LSAT, SAT, and TOEFL test prep and private tutoring. Professor Meissner received his Ph.D. in Management Science from Graduate School of Business at Columbia University (Columbia Business School) in New York City and is a recognized authority in the area of Supply Chain Management (SCM), Dynamic Pricing and Revenue Management. Currently, he holds the position of Full Professor of Supply Chain Management and Pricing Strategy at Kuehne Logistics University in Hamburg, Germany. Professor Meissner is a passionate and enthusiastic teacher. He believes that grasping an idea is only half of the fun; conveying it to others makes it whole. At his previous position at Lancaster University Management School, he taught the MBA Core course in Operations Management and originated three new MBA Electives: Advanced Decision Models, Supply Chain Management, and Revenue Management. He has also lectured at the University of Hamburg, the Leipzig Graduate School of Management (HHL), and the University of Mannheim. Professor Meissner offers a variety of Executive Education courses aimed at business professionals, managers, leaders, and executives who strive for professional and personal growth. He frequently advises companies ranging from Fortune 500 companies to emerging start-ups on various issues related to his research expertise. Please visit his academic homepage www.meiss.com for further information.

Manhattan Review Advantages

- **Time Efficiency and Cost Effectiveness**
  - The most limiting factor in test preparation for most people is time.
  - It takes significantly more teaching experience and techniques to prepare a student in less time.
  - Our preparation is tailored for busy professionals. We will teach you what you need to know in the least amount of time.

- **High-quality and dedicated instructors who are committed to helping every student reach her/his goals**

- **Manhattan Review's team members have combined wisdom of**
  - Academic achievements
  - MBA teaching experience at prestigious business schools in the US and UK
  - Career success

- **Our curriculum & proprietary Turbocharge Your GMAT course materials**
  - About 600 pages of well-illustrated and professionally presented strategies and exclusive problems for both the Verbal and the Quantitative Sections
  - 200+ pages of detailed solutions
  - 300-page of internally developed Quantitative and Verbal vocabulary list with detailed definitions, related words and sentence examples
  - Challenging Online CATs (Included in any course payments; Available for separate purchases)

- **Combine with Private Tutoring for an individually tailored study package**

- **Special Offer for Our Online Recording Library (Visit Online Library on our website)**

- **High-quality Career, MBA & College Advisory Full Service**

- **Our Pursuit of Excellence in All Areas of Our Service**

Visit us often at www.ManhattanReview.com.
(Select International Locations for your local content!)

International Phone Numbers & Official Manhattan Review Websites

Manhattan Headquarters	+1-212-316-2000	www.manhattanreview.com
USA & Canada	+1-800-246-4600	www.manhattanreview.com
Australia	+61-3-9001-6618	www.manhattanreview.com
Austria	+43-720-115-549	www.review.at
Belgium	+32-2-808-5163	www.manhattanreview.be
China	+86-20-2910-1913	www.manhattanreview.cn
Czech Republic	+1-212-316-2000	www.review.cz
France	+33-1-8488-4204	www.review.fr
Germany	+49-89-3803-8856	www.review.de
Greece	+1-212-316-2000	www.review.com.gr
Hong Kong	+852-5808-2704	www.review.hk
Hungary	+1-212-316-2000	www.review.co.hu
India	+1-212-316-2000	www.review.in
Indonesia	+1-212-316-2000	www.manhattanreview.com
Ireland	+1-212-316-2000	www.gmat.ie
Italy	+39-06-9338-7617	www.manhattanreview.it
Japan	+81-3-4589-5125	www.manhattanreview.jp
Malaysia	+1-212-316-2000	www.manhattanreview.com
Netherlands	+31-20-808-4399	www.manhattanreview.nl
Philippines	+1-212-316-2000	www.review.ph
Poland	+1-212-316-2000	www.review.pl
Portugal	+1-212-316-2000	www.review.pt
Russia	+1-212-316-2000	www.manhattanreview.ru
Singapore	+65-3158-2571	www.gmat.sg
South Africa	+1-212-316-2000	www.manhattanreview.co.za
South Korea	+1-212-316-2000	www.manhattanreview.kr
Sweden	+1-212-316-2000	www.gmat.se
Spain	+34-911-876-504	www.review.es
Switzerland	+41-435-080-991	www.review.ch
Taiwan	+1-212-316-2000	www.gmat.tw
Thailand	+66-6-0003-5529	www.manhattanreview.com
United Arab Emirates	+1-212-316-2000	www.manhattanreview.ae
United Kingdom	+44-20-7060-9800	www.manhattanreview.co.uk
Rest of World	+1-212-316-2000	www.manhattanreview.com

Contents

Chapter 1

Welcome

Dear Students,

Here at Manhattan Review, we constantly strive to provide you with the best educational content for standardized test preparation. We make a tremendous effort to keep making things better and better for you. This is especially important for an examination such as the SAT. As you know, from Spring'24 on, the SAT goes for a major change. The Digital SAT is challenging now.

There are umpteen numbers of books on math for Digital SAT preparation. What is so different about this book? The answer lies in its approach to dealing with the questions. The book is meant to develop your fundamentals. It does not have problems with the DSAT's four question types: Algebra, Advanced Math, Problem Solving & Data Analysis, and Geometry & Trigonometry. The purpose of the book is to sharpen your math fundamentals. We have tried our best to cover all the possible concepts in DSAT math.

The concepts are explained with the help of text and graphics. It is a treat to read the book along with relevant graphics; pictures speak louder than words!

A few short-cut techniques, such as the 'See-Saw technique' and 'Some Application of Percent and Fraction' on the topic of Fraction and Percent, are a copyright treat for the students. We believe that you will save your precious time by solving the questions with ease.

The Manhattan Review's "DSAT-Math Essentials" book is holistic and comprehensive in all respects. Should you have any queries, feel free to write to me at *info@manhattanreview.com.*

Happy Learning!

Professor Dr. Joern Meissner
& The Manhattan Review Team

Chapter 2

Introduction to The Digital SAT

In 2023–24, the College Board changed the previous SAT to a digital format. The digital version of the SAT is reduced to 2 hours, 14 minutes, as opposed to the previously 3 hours, 15 minutes. This change is welcome because it validates the relevance and significance of standardized testing.

When Can a Student Take the Digital SAT?

The College board publishes test dates for the digital SAT, as before.

What changes did the College Board make to the digital SAT?

In addition to reducing testing time to 2 hours, 14 minutes from the previous 3 hours, 15 minutes, the digital SAT will allow more time per question.

The following are the significant changes compared to the previous SAT content:

- Shorter reading passages with a single (discrete) question per text
- Texts from a broader range of college-level exams
- Calculator access for the entire math section
- The digital SAT scores are provided more quickly, allowing students and universities to make quicker college decisions.

Advantages of digital SAT

Safe: The paper-and-pencil version of the test could be canceled if there are problems outside or inside the school. The digital SAT is safer and better protected because each student will have a unique test form and will use their own or a test center-issued device.

Faster Score: Students get their scores in days as opposed to weeks.

What are the differences between the digital SAT and the previous SAT?

The digital SAT is a simpler version of the previous SAT. The new structure will place more emphasis on college-ready abilities and less on tenacity and speed. The previous SAT Reading section, with its long passages and frequently vague history passages, demanded patience to solve. The digital SAT incorporates texts from the humanities and sciences that dive into more modern topics in finance, economics, and the environment, as well as a greater variety of short passages.

Also, there is relatively greater emphasis on using both textual and graphic content to answer questions.

Since calculators can be used for the entire math section, it seems like the test has been changed in a big way. In addition to the focus on algebra and word problems on the SAT, there may be a shift to more straightforward questions, like those on the ACT, and more geometry and trigonometry questions.

As the College Board changed test content to make it better for digital distribution, the digital SAT still tests the same skills and knowledge learned in high school that are most important for college and job readiness.

Digital SAT Vs. Previous SAT: What Has Not Changed?

Despite the fact that the digital SAT comes up with a number of changes that is beneficial to both students and colleges, many of the SAT's most essential components stay unaltered.

- The digital SAT still tests students on the skills and knowledge they need to be ready for college and careers.
- The SAT is scored on the same 1600-point scale, so teachers and students are able to see how much they have improved over time.
- The digital SAT cannot be taken at home but rather in a school or testing center in a proctored setting.

Does a student Need a Device to Take the Digital SAT?

The digital SAT is taken on a laptop or tablet. Students can use their personal or test center-issued devices. If a student does not have a laptop, they may request one from the College Board, and it is provided to them on their SAT Day. If a student has any device or connectivity issues on the test day, the College Board has designed the test application so that it survives internet outages. Students are able to continue with the test without interruption if the internet goes down; their responses are stored, and they will not lose any testing time.

How Challenging Is It to Take the SAT Online?

The good news is that most students who took the digital SAT found the new version of the test easier to understand and use. When the College Board says that the digital version of the test is more "intuitive," it implies that it is "easier to navigate" and "easier to answer without becoming exhausted."

The digital SAT interface has a provision for a clock, which makes it easier for students to keep track of time.

In addition to highlighters, strikethrough, and bookmarks for review, there is an integrated calculator and a provision for a standard math formula sheet. At a single location, test-takers find all the resources necessary to answer questions.

What Tools Are Available for Students Taking the Digital SAT?

The test application includes many testing tools. Here are two examples:

- You can mark questions for review using the bookmark feature.
- Test-takers have the option to display or hide the digital clock placed at the top of their testing screen.
- During the math section, students may use the integrated graphing calculator (or one can bring their own basic calculator).
- A reference sheet with standard mathematical formulas for each question is provided.

When and How to Study for the Digital SAT?

Start preparing for the digital SAT as quickly as possible. Focus on making noticeable gains in every area: reading, writing, and math. Don't be hesitant to monitor your progress using the previous SAT tests. There is no need to take full-length tests, but you may always attempt to earn a perfect score on practice tests. Before starting to read, emphasize the need to develop comprehension skills and knowledge of previous global issues and concerns. This will gradually boost your vocabulary and reading speed, preparing you for future SAT reading success.

How to Study for the Digital SAT?

Reading, writing, grammar, and math skills are still the most important parts of the SAT for measuring college readiness.

This is how to begin preparing for each section of the digital SAT:

- **Reading and Writing (RW):** Start with a variety of humanities- and science-related texts. You should read the New York Times, the Economist, the New Scientist, and the Washington Post. You should start reading the editorials. Learn about contemporary environmental and geopolitical concerns. Focus on being a knowledgeable reader in general. Instead of being a passive recipient of information, develop views. Application of grammar and punctuation rules is still tested on the SAT.

- **Math**: Start practicing questions on all the math topics on the previous SAT. In addition, pay close attention to the geometry and trigonometry questions on the official SAT material and Manhattan Review guides. Learn the formulae and their application to the previous SAT topics, as well as the rationale behind their operation.

How Would These Changes Impact College Admissions?

Since the SAT's basic content, level of difficulty, and scoring system did not change much, we expect schools to evaluate the digital version in the same way they evaluate the previous version. In addition, colleges make an attempt not to fault students for situations beyond their control, so taking one version of the SAT will not negatively affect your college applications relative to those who took the other version.

Even with these changes, colleges have a very different view of the SAT and ACT than they had a few years ago. Several schools are paying less attention to standardized test scores and more attention to other parts of an application, such as a student's grade point average (GPA), the courses they took in high school, and extracurricular activities. The COVID-19 epidemic made this worse by making it difficult or impossible for many students to take standardized tests for several months. Almost every school made testing optional in 2020 and 2021, and hundreds of colleges have decided to stay test-optional permanently. If a school is test-optional, you have the choice of submitting SAT or ACT scores with your application. If you do not, the other application aspects will carry more weight.

The College Board is aware of these trends, and many of the changes to the SAT were made to make it easier for students who might have been turned off by the length and difficulty of the test. Still, a College Board survey showed that 83% of students wanted to be able to send test scores with their college applications, and almost all schools still let them, so the SAT is here to stay.

Key Takeaways

The following are the six most significant takeaways:

- The SAT is administered digitally.
- The duration of the SAT is reduced (to about 2 hours instead of about 3 hours).
- Students have more time to answer each question.
- Calculator is permitted for the entire math section of the test.
- Reading passages is of shorter length. Each passage has a single question.

- Students get their digital SAT scores within days rather than weeks.

Digital SAT Math content

Content	**Topics**	**Number of questions**
Algebra	Fundamental concepts used in Algebra; arranging formulae, linear equations, inequalities, etc.	13–15
Advanced Math	Advanced concepts in Algebra, incl. absolute value, quadratic, exponential, polynomial, rational, radical, and other nonlinear equations	13–15
Problem Solving & Data Analysis	Understanding qualitative & quantitative data, analyzing relationship, incl. graphical; Ratio & Proportion, Percents, and units of measurements	5–7
Geometry & Trigonometry	Geometry (angles, triangles, area, and volume), Trigonometry (Trigonometric ratios)	5–7
	Total time (70 minutes)	**44**

Chapter 3

DSAT—Math

3.1 Understanding numbers

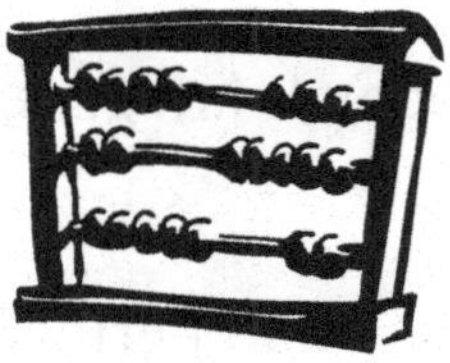

Numbers—We all are familiar with numbers, but the devil lies in the detail!

Let us understand numbers from the SAT perspective.

3.1.1 Classification of numbers

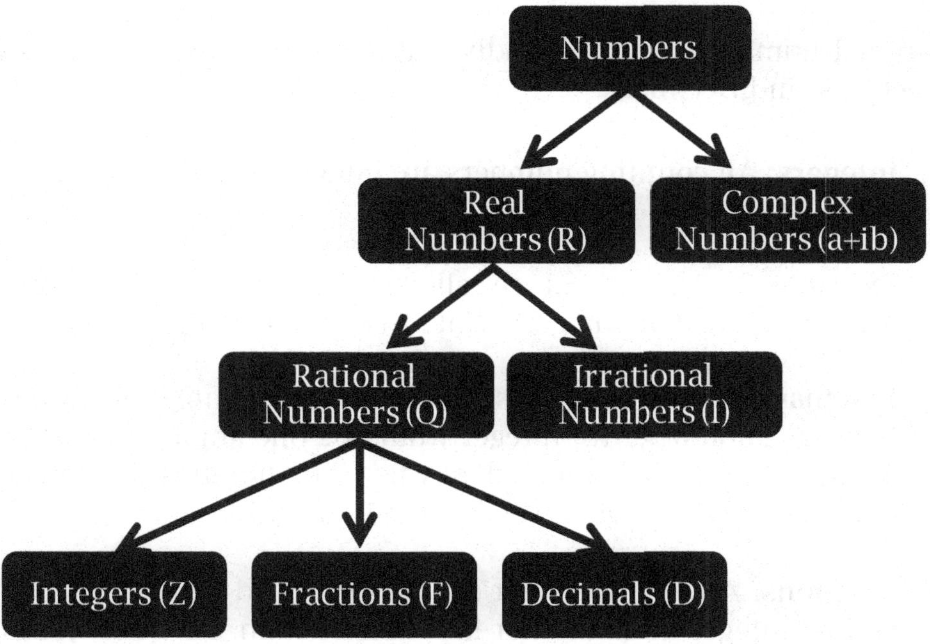

(A) **Real numbers:** Almost all the numbers you can imagine are Real numbers. Any number that can be represented on a number line is a real number. These include $0, 1, -3, 4/3,$

$-0.3457, 48.87, \sqrt{3}, \pi$, and many more!

If most numbers are real numbers, then what are not real numbers?

The numbers such as $\sqrt{-1}$, $(2+\sqrt{-5})$, etc. are not real numbers. They are Imaginary number and Complex numbers respectively.

Real numbers can further be classified into Rational and Irrational numbers.

(a) **Rational numbers:** Numbers that can be expressed in the form of $\frac{p}{q}$, where p and q are integers, and $q \neq 0$. Remember that p may be 0.

The above definition may look a little scary to those who are not very familiar with the concepts of numbers; however the good news for them is that you need not know the above definition to master the SAT math.

So what do rational numbers include? Simply put, these include all integers, decimal, and fractions.

So what do rational numbers exclude? These exclude roots of non-perfect square numbers, roots of non-perfect cube numbers, special numbers such as $\pi = 3.14, e = 2.718$, and many others, basically, non terminating & non recurring decimals.

Rational numbers can be broadly classified into three categories—Integers, Fractions, and Decimals.

i. **Integers:** All counting numbers are integers whether negative, positive or zero.

Example: {1, 2, 3, 4, 5, ...} are called positive integers; $\{-1, -2, -3, -4, -5, ...\}$ are called negative integers. Note that "0" is also an integer.

You may come across terms like non-negative integers or non-positive integers. A non-negative integer would be one among the set of {1, 2, 3, 4, 5, ...}; note that "0" is included; whereas a non-positive integer would be one among the set of $\{-1, -2, -3, -4, -5, ...\}$.

ii. **Fractions:** Any number that can be expressed with an integer numerator and an integer denominator is called fraction. In other words, a number that can be expressed in the form of x/y is a fraction such that $y \neq 0$. **Example:** $1/3, 5/4, -3/8$, etc.

iii. **Decimal numbers:** Decimal numbers are another way of expressing fractions. The decimal numbers are written with the use of a decimal (".").

The left of the decimal point (.) has place values of digits for units, tens, hundreds, thousands, and more; whereas the right of the decimal point (.) has place values of digits for tenths, hundredths, thousandths, and more.

Only those decimals which are 'Terminating,' for example $2.4, 4.35, -2.1$, etc. or 'Recurring,' for example $2.3333, -4.151515..., 2.3404040$, etc. are Rational.

This is because only these can be expressed in the form p/q, as discussed earlier. For example: A terminating decimal. 2.35 is simply $235/100 = 47/20$, and is recurring. A recurring decimal, say 2.343434... can be analyzed as follows:

Say $x = 2.343434...$
$\Rightarrow 100x = 234.343434$
Subtracting: $100x - x = 99x = 234.343434 - 2.343434 = 232$
$\Rightarrow x = \frac{232}{99}$, and hence is a rational number.

We will discuss place value in the following pages.

(b) **Irrational numbers:** Numbers that cannot be expressed in the form of $\frac{p}{q}$, where p and q are integers, and $q \neq 0$ are irrational numbers or in other words, all real numbers that are not rational are irrational numbers. As stated earlier that these include roots of non-perfect squared numbers: 2, 5 and others; roots of non-perfect cube numbers, special numbers such as $\pi = 3.14, e = 2.718$, and many others.

By default any number mentioned in the SAT is a **real number**; so, you must NOT assume that it is an integer unless stated as such.

(B) **Complex numbers:** As stated earlier that the numbers such as $\sqrt{-1}, (2 + \sqrt{-5})$, etc. are not real numbers. They are Imaginary number and Complex numbers, respectively. Fortunately, the College Board has excluded them from the SAT.

3.1.2 Number line

A number line is a line with "0" as its center. Numbers on the right side of "0" are positive and those on the left side are negative. Number line helps define the direction of measurements.

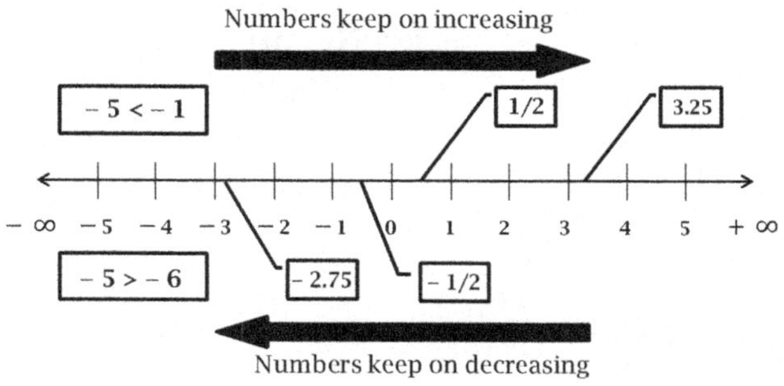

3.1.3 Basic operations on numbers

Symbol	Mathematical operation	Application	Examples
$+$	Addition	Sum, Add, Plus, Total, Increase	• $2 + 3 = 5$; (Two positives will be added) • $(-2) + (-3) = -5$; (Two negatives will be added) • $2 + (-3) = -1$; (A positive & a negative will be subtracted) • $(-2) + 3 = 1$; (A negative and a positive will be subtracted)
$-$	Subtraction	Deduct, Minus, Decrease, Less, Reduce, Difference	• $2 - 3 = -1$; (Two positives will be subtracted) • $(-2) - (-3) = 1$; (Two negatives will be subtracted) • $2 - (-3) = 5$; (A positive & a negative will be added) • $(-2) - 3 = -5$; (A negative & a positive will be added)
$\times$	Multiplication	Multiply, Product, Times, $\times$, By	• $2 \times 3 = 6$; (Positive $\times$ Positive = Positive) • $(-2) \times (-3) = 6$; (Negative $\times$ Negative = Positive) • $2 \times (-3) = -6$; (Positive $\times$ Negative = Negative) • $(-2) \times 3 = -6$; (Negative $\times$ Positive = Negative)

÷	Division	Divide, Quotient, Part of, Per unit calculation	• $6 \div 3 = 2$; (Positive ÷ Positive = Positive) • $(-6) \div (-3) = 2$; (Negative ÷ Negative = Positive) • $6 \div (-3) = -2$; (Positive ÷ Negative = Negative) • $(-6) \div 3 = -2$; (Negative ÷ Positive = Negative)

3.1.4 Face value and place value

The number system we use is set up from ten symbols only $\{0, 1, 2, 3, 4, 5, 6, 7, 8, 9\}$. Each symbol is called a digit and each has its value called **Face value.** So, the face value of "9" is nine times the face value of the digit "1" or three times the face value of the digit "3." For example, for numbers 243 & 234, the face values of digits "2," "3," and "4" do not vary upon their placement.

Place value of a digit varies depending on where it is placed in a number. For example, for numbers 234, and 243, the place values of digits "3," and "4" are different.

If a digit placed on the left of another digit, the place value of the first digit is 10 times of the other digit. For a number 234, the place value of digit "2" is 200, that of digit "3" is 30, and that of "4" is 4. Similarly, the place values of digits placed after the decimal point (·) reduces in its value by 1/10 times. This can be better understood this way.

- **$243 = 2 \times 100 + 4 \times 10 + 3 \times 1$**
- **$234 = 2 \times 100 + 3 \times 10 + 4 \times 1$**

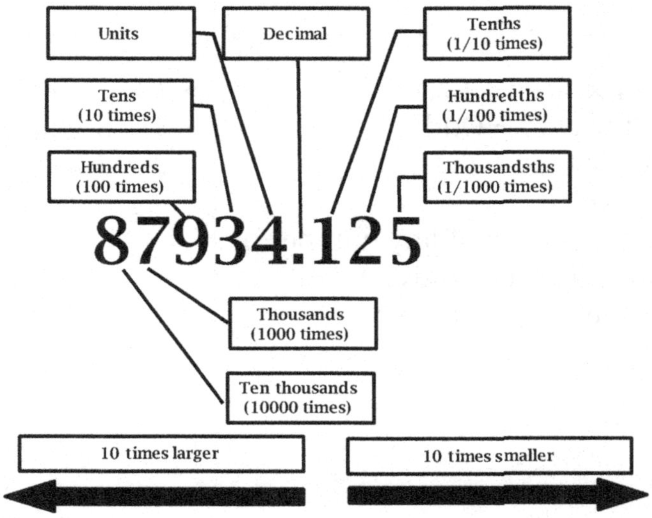

3.1.5 Even and Odd numbers

Even number: An integer completely divisible by "2," without leaving a remainder is called an **Even number.** In other words, an even number can be expressed as "$2m$," where "m" is an integer. As per this definition, "0" is an even integer. **Example:** $\{0, 2, 4, 6, 8, ...\}, \{-8, -4, 10, 18, ...\}$.

Odd number: An integer when divided by "2" leaves a remainder "1" is called **Odd number.** In other words, an odd number can be expressed as $(2m + 1)$, where "m" is an integer. All the integers are either even or odd integers. **Example:** $\{1, 3, 5, ...\}, \{-3, -1, 9, ...\}$.

Basic operations on Even and Odd numbers:

Mathematical operation	**Application**	**Examples**
Addition/ Subtraction	Even ± Even = Even; Odd ± Odd = Even; Even ± Odd = Odd Any number of even numbers can make an even number; however only an even number of odd numbers can make an even number; e.g.: Two odd numbers on addition give even, four odd numbers on addition also give even; however three odd numbers on addition give odd number.	• 4 + 2 = 6; 4 - 2 = 2; Even • 7 + 3 = 10; 3 - 7 = - 4; Even • 5 + 2 = 7; 5 - 2 = 3; Odd
Multiplication	Even × Even = Even; Odd × Odd = Odd; Even × Odd = Even Multiplication of any number of odd numbers always results in another odd number.	• 4 × 2 = 8; Even • 3 × 3 = 9; Odd • 2 × 3 = 6; Even

	If the product of two integers is Even and one of them is an Even integer, we cannot comment on the nature of the other number, i.e. whether it is Even or Odd.	
Division	Not every operation of Division is completely divisible. We show you the results if the division leaves no remainder. Odd ÷ Odd = Odd; Even ÷ Odd = Even; Odd ÷ Even = Not divisible; Even ÷ Even = Even or Odd	• 9 ÷ 3 = 3; Odd • 10 ÷ 5 = 2; Even • 11 ÷ 2 = ? Not divisible • 4 ÷2 = 2; Even • 10 ÷ 2 = 5; Odd

3.1.6 Consecutive numbers

An ordered set of continuous integers written in ascending order are called **Consecutive numbers.**

Example: $\{23, 24, 25\}, \{-2, -1, 0, 1, 2, 3\}$

Example of consecutive Even numbers: $\{12, 14, 16\}$

Example of consecutive Odd numbers: $\{11, 13, 15\}$

3.1.7 Prime numbers

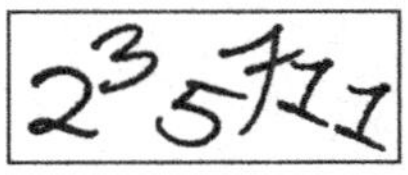

A number greater than "1" which has no factors other than "1" and the number itself, is called **Prime number**. A number with exactly two distinct factors is called prime. Hence, "1" cannot be considered as prime since it has only one factor, "1" itself.

Example: $\{2, 3, 5, 7, 11, 13, 17, 19, 23, 29, ...\}$. There are infinitely many Prime numbers.

First 25 prime numbers are:

$\{2, 3, 5, 7, 11, 13, 17, 19, 23, 29, 31, 37, 41, 43, 47, 53, 59, 61, 67, 71, 73, 79, 83, 89, 97\}$

"2" is the only even prime number or all the prime numbers are odd except "2."

Also, the difference between any two primes greater than "2" is always even.

Another point to note is that any prime, other than "2" or "3," if divided by "6," will always leave a remainder of "1" or "5" (= -1). Thus, any prime can be expressed as "$6k \pm 1$," where k is some positive positive integer.

However, you should note that not all numbers which leave a remainder of 1 or 5 when divided by 6 are prime, for example, 25, 35, etc.

How to verify whether a number "n" is a prime number:

Step1: Take square root of the number; if square root is an integer the number is not prime number, else follow the next step.

Step2: Say the largest integer closest to the square root of the number "n" is "m"; where $m < \sqrt{n}$;

Step3: Divide "n" by all the prime numbers from "2 to m," inclusive: if n is divisible by any number among them, the number "n" is not prime, else prime

=> Let us take a number "$n = 101$."

Step1: $10 < \sqrt{101} < 11$; not an integer;

Step2: $\sqrt{101} =\approx 10 = m$;

Step3: Divide 101 by prime numbers from 2 to 10 i.e. 2, 3, 5, & 7. Since 101 is not divisible by any prime number among these, hence 101 is a prime number.

=> Let us take a number "$n = 87$."

Step1: $9 < \sqrt{87}, 10$; not an integer;

Step2: $\sqrt{87} =\approx 9 = m$;

Step3: Divide 87 by prime numbers from 2 to 9 i.e. 2, 3, 5, & 7. Since 87 is divisible by 3, hence 87 is not a prime number

3.1.8 Co-Prime numbers

Two numbers are co-prime to each other if they have no factor common between them except '1'.

Example: 14 & 15: The factors of "14" are {1, 2, 7, 14}, and the factors of "15" are {1, 3, 5, 15}. There is no factor common between "14 & 15" except '1,' so "14 & 15" are co-prime to each other.

It is important to know that the LCM of co-prime numbers is their product, and GCD or HCF is "1." Say two numbers "a" and "b" are co-prime to each other, then **LCM** = $a \times b$, and **GCD = 1**.

Note that any two consecutive numbers are always co-prime to each other. Also 1 is co-prime to any number.

3.1.9 Rounding off

Rounding off a number means to get a significant and practically usable number. Say, the distance between Manhattan to JFK airport is 15.24 miles; we can truncate the hundredth place digit to get a number which is rounded off to tenth place equalling 15.2 miles. However, by doing this, we do get a less accurate number. The degree of rounding off a number depends on how much accurate number you want. In this case, even the additional 0.2 miles may seem insignificant to a regular commuter. So, 15.2 can further be rounded off to the nearest tens as 15 miles which is further less accurate.

CASE I:

If the dropped digit is less than '5' (i.e. 0, 1, 2, 3, 4), the digit to the left of the dropped digit should be unchanged. For example, 5.312 rounded to the tenths digit is 5.3.

CASE II:

If the dropped digit is greater than '5' (i.e. 6, 7, 8 or 9), the digit to the left of the dropped digit should be increased by '1'. For example, 5.326 rounded to the hundredths digit is 5.33, but, rounded to the tenths digit is 5.3; similarly, 4.961 becomes 5.00, if rounded to the nearest integer.

CASE III:

If the dropped digit is '5,' and all the figures following the '5' are '0,' or if there is no digit after the '5':

- The digit to the left of the dropped digit should be unchanged, if that last figure is even, and
- The digit to the left of the dropped digit should be increased by 1 if that last figure is odd. For example, 3.195 becomes 3.20, whereas 2.365 becomes 2.36, both when rounded to the hundredths digit.

CASE IV:

If the dropped digit is '5,' and all the figures following the '5' are not all '0,' the digit to the left of the dropped digit should be increased by '1'. For example, 4.7501 rounded to the hundredths digit is 4.8.

The same rules for rounding can be applied in the following cases as well:

- 2,134 rounded to the nearest tens is 2,130
- 2,139 rounded to the nearest tens is 2,140
- 2,135 rounded to the nearest tens is 2,140
- 2,145 rounded to the nearest tens is 2,150
- 2,145.1 rounded to the nearest tens is 2,150
- 2,145 rounded to the nearest hundreds is 2100
- 2,155 rounded to the nearest hundreds is 2,200
- 2,168 rounded to the nearest hundreds is 2,200
- 2,134 rounded to the nearest hundreds is 2,100

Note that the number, rounded off, can be greater than or less than the original number. 15.8454 rounded off to $15.8 < 15.8454$, whereas another number 15.8554 rounded off to $15.9 > 15.8554$.

3.1.10 Order of operations—PEMDAS

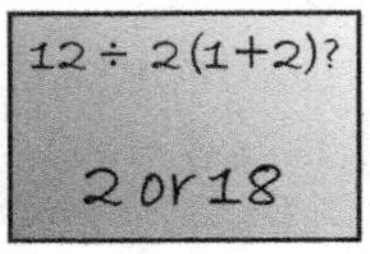

What is the value of $2\{10^2 - 2(6 \times 5^2 - 50 + 25 \div 5)\}$?

Follow the order of operation given as below:

P: Parentheses first

E: Exponents (i.e. Powers and Square Roots, etc.)

MD: Multiplication and Division (left-to-right)

AS: Addition and Subtraction (left-to-right)

Expression	**Comments**
$2\{10^2-2(6\times5^2-50+25\div5)\}$	Solve the parenthesis first. If there are more than two parentheses, follow the order: [ ], { }, & ( ). Here, we will solve { } first, and then ( ). **(P)**
$=2\{\mathbf{100}-2(6\times\mathbf{25}-50\div25)\}$	Among the operators exponent, multiplication, division, addition, subtraction, solve the exponent first. **(E)**
$=2\{100-2(\mathbf{150}-50+5)\}$	Multiplication and division can be done simultaneously. **(MD)**
$=2\{100-2(\mathbf{105})\}$ $=2\{100-\mathbf{210}\}$ $=2\{-110\}$ $=-220$	Addition and subtraction can be done simultaneously. **(AS)**

3.1.11 Divisibility

A number is said to be divisible by another number if the division does not leave any remainder. Following are quick reckoner rules for divisibility test.

Divisor	Condition: A number is divisible	Example
2	if the unit digit is Even: {0, 2, 4, 6, 8}	26**8**; 126,59**6**
3	if the Sum of all the digits is divisible by "3"	21,681; Sum of digits = 2+1+6+8+1 = 18; 18 div. by 3 implies that 21,681 is divisible by 3. 89,896; Sum of digits = 8+9+8+9+6 = 40; 40 not div. by 3 implies that 89,896 is not divisible by 3.
4	if the number formed out of last two digits is divisible by "4"	1,2**24** and 2,3**56** are div. by 4 as 24 and 56 are divisible by 4. 1,2**54** and 2,569,8**66** are not div. by 4 as 54 and 66 are not divisible by 4.
5	if the unit digit is "0" or "5."	1,22**0** and 2,35**5** are div. by 5. 1,25**4** and 2,569,86**6** are not div. by 5.
6	if the number is divisible by both "2" and "3"; follow the divisibility rules of "2" and "3."	1254, and 89,892 are div. by 6.
8	if the number formed out of the last three digits is divisible by "8."	891,**224** and 2,542,**352** are div. by 8. 561,**354** and 2,569,**866** are not div. by 8.

9	if the Sum of all digits of the number is divisible by "9."	21,681; Sum of digits = 2+1+6+8+1 = 18; 18 div. by 9 implies that 21,681 is divisible by 9. 89,896; Sum of digits = 8+9+8+9+6 = 40; 40 not div. by 9 implies that 89,896 is not divisible by 9.
10	if the unit digit is "0."	12,340 and 5,245,870.

3.1.12 Factors

All the possible divisors of an integer which leave no reminder are called factors.
Example: The factors of '6' are 1, 2, 3, & 6. We can write 6 as 1×6 or 2×3.

Factors of 60:

$$\begin{aligned} 60 &= 1 \times 60; \\ &= 2 \times 30; \\ &= 3 \times 20; \\ &= 4 \times 15; \\ &= 5 \times 12; \\ &= 6 \times 10. \end{aligned}$$

So, the factors of '60' are $\{1, 2, 3, 4, 5, 6, 10, 12, 15, 20, 30, 60\}$. Remember that '1' and the number itself are also the factors of a number.

Number of factors

Say, a number N can be written as $\boldsymbol{N = a^x b^y c^z}$, where a, b, and c are prime factors of the number N, and x, y, and z are positive integers, then

=> **Number of factors** (including '1' and the number itself) = $(x+1)(y+1)(z+1)$;

=> **Number of factors** (excluding '1' and the number itself) = $(x+1)(y+1)(z+1) - 2$.

Number of factors of 24:

$24 = 2 \times 2 \times 2 \times 3 = 2^3 \times 3^1$; here $a = 2$ & $b = 3$ are prime numbers, and $x = 2$ & $y = 1$.

Hence the number of factors of 24 including '1' and the number itself
$= (3+1)(1+1) = 4 \times 2 = 8$.

Let us verify this.

Factors of 24 are: $\{1, 2, 3, 4, 6, 8, 12, 24\}$; total number of factors equals to 8.

Let us take one more example.

Number of factors of 2,520:

$2,520 = 2 \times 2 \times 2 \times 3 \times 3 \times 5 \times 7 = 2^3 \times 3^2 \times 5 \times 7$; here $2, 3, 5, \&\ 7$ are prime numbers.

Hence the number of factors of 2,520 including '1' and the number itself
$= (3+1)(2+1)(1+1)(1+1) = 4 \times 3 \times 2 \times 2 = 48$.

Sum of all the factors:

Say, a number N can be written as $N = a^x b^y c^z$, where a, b, and c are prime factors of number N, x, y, and z are positive integers, then

Sum of all the factors (including 1 and the number itself)

$$= \left[\frac{a^{x+1} - 1}{a - 1}\right]\left[\frac{b^{y+1} - 1}{b - 1}\right]\left[\frac{c^{z+1} - 1}{c - 1}\right]$$

Sum of all the factors of 60:

$60 = 2^2 \times 3^1 \times 5^1;$

Sum of all the factors (including 1 and the number itself)

$$= \left[\frac{2^{2+1} - 1}{2 - 1}\right]\left[\frac{3^{1+1} - 1}{3 - 1}\right]\left[\frac{5^{1+1} - 1}{5 - 1}\right]$$

$= 7 \times (8/2) \times (24/4) = 168.$

Let us cross check this. Sum of all the factors of $60 = 1 + 2 + 3 + 4 + 5 + 6 + 10 + 12 + 15 + 20 + 30 + 60 = 168$.

3.1.13 Perfect Square numbers

A perfect square number is a non-negative integer that can be expressed as the product of an integer with itself. In other words, a number which is a square of any integer is called Perfect square number. Perfect square numbers are always non-negative integers.

Example: $\{0, 1, 4, 9, 16, 25, 36, 49, ...\}$. Some mathematicians do not consider that "0" is a perfect square number. Moreover the SAT does not assume that you know the meaning of "perfect square number." The phrase used typically is "square of an integer."

We can see that $1 = 1 \times 1 = -1 \times -1$; and $4 = 2 \times 2 = -2 \times -2$.

3.1.14 Multiples

While the factors divide the number completely, multiples are completely divisible by the number itself.

Example: The factors of '6' are 1, 2, 3, & 6; and the multiples of '6' are 6,12,18,24,30,...

=> A number is both a factor and a multiple of itself.

=> There are a finite number of factors of a number, but there are an infinite number of multiples.

In other words, we can write that the **multiples of a number $N = k \times N$,** where k is a positive integer. So the multiples of a number, say, xy^2 would be $xy^2, 2xy^2, 3xy^2, 4xy^2, \ldots$ and the factors of xy^2 would be $\{1, x, xy, y^2, xy^2\}$.

3.1.15 Lowest Common Multiple—LCM

We all know what the LCM is; but occasionally we may get confused with HCF.

The LCM has three properties:

(1) It is a **multiple;** for numbers 6 & 10, the multiples of 6 are $\{6, 12, 18, 24, 30, 36, \ldots\}$, and the multiples of 10 are $\{10, 20, 30, 40, 50, 60, 70, 80, 90, \ldots\}$, so LCM is one among these multiples.

(2) It is a **common multiple** among all the multiples of 6, & 10; for numbers 6 & 10, the common multiples are $\{30, 60, 90, \ldots.\}$, so it is one among these multiples.

(3) It is the **least common multiple (LCM)** among all the common multiples; for numbers 6 & 10, the least common multiple is **30: LCM.**

So, in other words, LCM of numbers is the smallest possible number which gets divided by each number without leaving any reminder.

Did you notice that in math, we never come across a term—Highest Common Multiple? It is because it is insignificant to discuss this as the Highest Common Multiple for any set of numbers would always be infinite; multiples of numbers are never ending.

Computation of LCM by factor method:

Example: Computation of LCM of 24, 54, and 70:

Step1: Factorize the numbers into prime factors.

=> $24 = \mathbf{2^3} \times 3$;
=> $54 = 2 \times \mathbf{3^3}$;
=> $70 = 2 \times \mathbf{5} \times \mathbf{7}$

Step2: Select the term with the highest exponent for each prime factor.

=> The prime factors with their highest exponents are $2^3, 3^3, 5,$ & 7.

Step3: Multiply the numbers selected in step 2; it is the LCM of numbers.

=> **LCM** = $2^3 \times 3^3 \times 5 \times 7 =$ **7,560.**

Computation of LCM of algebraic expressions:

Example: Computation of LCM of $10xy^2z^5, 15x^2y^3z^4$, & $24x^3y^3z$:

=> $10xy^2z^5 = 2 \times \mathbf{5} \times x \times y^2 \times \boldsymbol{z^5}$;
=> $15x^2y^3z^4 = \mathbf{3} \times 5 \times x^2 \times \boldsymbol{y^3} \times x^4$;
=> $24x^3y^3z = \mathbf{2^3} \times 3 \times \boldsymbol{x^3} \times y^3 \times z$

LCM of $10xy^2z^5, 15x^2y^3z^4$, & $24x^3y^3z = 2^3 \times 3 \times 5 \times x^3 \times y^3 \times z^5 = \mathbf{120}\boldsymbol{x^3y^3z^5}$

3.1.16 Highest Common Factor—HCF OR Greatest Common Divisor—GCD

Like LCM, the HCF/GCD also has three properties:

(1) It is a **factor;** for numbers 18 & 24, the factors of 6 are $\{1, 2, 3, 6, 9, 18\}$, and the factors of 24 are $\{1, 2, 3, 4, 6, 8, 12, 24\}$, so HCF is one among these factors.

(2) It is a **common factor** among these factors; for numbers 18 & 24, the common factors are $\{1, 2, 3, 6\}$, so it is one among these factors.

(3) It is the **highest common factor (HCF)** among all the factors; for numbers 18 & 24, the highest common factor is **6: HCF.**

So, in other words, HCF of numbers is the largest possible number which divides each number without leaving any reminder.

Did you notice that in math, we never come across a term—Lowest Common Factor? It is because it is insignificant to discuss this as the Lowest Common Factor for any set of numbers would always be '1.'

Computation of HCF by factor method:

Example: Computation of HCF of 24, 54, and 210.

Step1: Factorize the numbers into prime factors.

=> $24 = 2^3 \times 3$;
=> $54 = 2 \times 3^3$;
=> $210 = 2 \times 3 \times 5 \times 7$.

Step2: Select the term with the lowest common exponent for each **common** prime factor.

=> All the common prime factors with their lowest exponents are 2, & 3.

Step3: Multiply the numbers selected in step 2. It is the HCF/GCD of numbers.

=> HCF = $2 \times 3 = 6$.

Computation of HCF for algebraic expressions:

Example: Computation of HCF of $10xy^2z^5, 15x^2y^3z^4$, & $24x^3y^3z$:

=> $10xy^2z^5 = 2 \times 5 \times \boldsymbol{x} \times \boldsymbol{y^2} \times z^5 = 2 \times 5 \times (\boldsymbol{x}) \times (\boldsymbol{y^2}) \times (\boldsymbol{z}) \times z^4$;
=> $15x^2y^3z^4 = 3 \times 5 \times x^2 \times y^3 \times z^4 = 3 \times 5 \times (\boldsymbol{x}) \times x \times (\boldsymbol{y^2}) \times y \times (\boldsymbol{z}) \times z^3$;
=> $24x^3y^3z = 2^3 \times 3 \times x^3 \times y^3 \times \boldsymbol{z} = 2^3 \times 3 \times (\boldsymbol{x}) \times x^2 \times (\boldsymbol{y^2}) \times y \times (\boldsymbol{z})$

HCF of $10xy^2z^5, 15x^2y^3z^4$, & $24x^3y^3z = \boldsymbol{xy^2z}$

LCM and HCF revisited:

Numbers: N_1, N_2, N_3	**LCM of** $N_1, N_2, N_3 = \boldsymbol{120x^3y^3z^5}$ LCM is completely divisible by numbers	**HCF of** $N_1, N_2, N_3 = \boldsymbol{xy^2z}$ Numbers are completely divisible by HCF
$N_1 = 10xy^2z^5$	$\frac{LCM}{N_1} = \frac{120x^3y^3z^5}{10xy^2z^5} = 12x^2y$	$\frac{N_1}{HCF} = \frac{10xy^2z^5}{xy^2z} = 10z^4$
$N_2 = 15x^2y^3z^4$	$\frac{LCM}{N_2} = \frac{120x^3y^3z^5}{15x^2y^3z^4} = 8xz$	$\frac{N_2}{HCF} = \frac{15x^2y^3z^4}{xy^2z} = 15xyz^3$
$N_3 = 24x^3y^3z$	$\frac{LCM}{N_3} = \frac{120x^3y^3z^5}{24x^3y^3z} = 5z^4$	$\frac{N_3}{HCF} = \frac{24x^3y^3z}{xy^2z} = 24x^2y^2$

3.2 Exponents

Exponents, also called indices, are powers given to bases. Exponents tell the number of times a number is multiplied by itself.
Example: $2^4 = 2 \times 2 \times 2 \times 2 = 16$. Here '2' is the base and '4' is the exponent or index.

Here are some basic operations of indices:

Operations	Examples
$a^m \times a^n = a^{m+n}$	$2^3 \times 2^5 = 2^{3+5} = 2^8$
$(ab)^m = a^m \times b^m$	$(2 \times 3)^4 = 2^4 \times 3^4$
$a^m \div a^n = a^{m-n}$	$2^3 \div 2^5 = 2^{3-5} = 2^{-2}$
$(a^m)^n = a^{mn}$	$(2^2)^3 = 2^{2.3} = 2^6$
$\dfrac{1}{a^m} = a^{-m}$	$\dfrac{1}{2^3} = 2^{-3}$
$\sqrt[n]{a^m} = (a^m)^{1/n} = a^{m/n}$	$\sqrt[3]{2^6} = (2^6)^{1/3} = 2^{6/3} = 2^2 = 4$
$a^0 = 1$, if $a \neq 0$	$2^0 = 1$
$(-a)^{\text{Even}} = +(a)^{\text{Even}}$; a = positive number	$(-2)^4 = -2 \times -2 \times -2 \times -2 = 16$
$(-a)^{\text{Odd}} = -(a)^{\text{Odd}}$; a = positive number	$(-2)^3 = -2 \times -2 \times -2 = -8$ or $-(2)^3 = -(2 \times 2 \times 2) = -8$

Note that $(a^m)^n \neq a^{m^n}$

- $(2^2)^3 = 2^{2.3} = 2^6 = 64$; • $2^{3^2} = 2^9 = 512 \neq 64$

3.2.1 Roots and Surds

Root is a familiar word; you have heard this before—square root, cube root, etc.

Square root:

It is opposite of square. While square of a number means multiplying a number with itself to get a new number, "Square root" means to get a number such that multiplying it with itself would give the original number.

Example: Say a number is $x = 4$, then the 'square $= x^2 = 4^2 = 16$,' and 'square root $= \sqrt[2]{x} = \sqrt[2]{4} = 2; 2 \times 2 = 4$'. Some consider that $\sqrt[2]{4} = 2$ or -2. However, it is not so. Its value is +2 only.

The confusion starts from here:

If $x^2 = 16$, what is x?

While taking the square root of both the sides, we get $\sqrt[2]{x^2} = \sqrt[2]{16} => \pm x = 4 => x = \pm 4$. Note that $\sqrt[2]{x} = 2$, not ± 2.

Remember that **square** of a number is NOT always greater than its **square root.** If a number is $x = 4$, then the '$x^2 = 16$,' and $\sqrt{x} = 2; 16 > 2$; however if a number is $x = 1/4$, then the '$x^2 = 1/16$,' and '$\sqrt{x} = 1/2$'; $1/16 < 1/2$.

Square of a negative number is possible, but the **square root** is not possible, it would be an imaginary number. If a number is $x = -4$, then '$x^2 = 16$,' and '$\sqrt{x} = \sqrt[2]{-2}$;' an imaginary number.

Cube root:

Like square root, which is opposite of square, cube root is opposite of cube. Cube of a number means multiplying a number with itself thrice to get a new number, "Cube root" means to get a number such that multiplying it with itself thrice would give the original number.

Example: Say a number is $x = 8$, then the 'cube $= x^3 = 8^3 = 512$,' and 'cube root $= \sqrt[3]{x} = \sqrt[3]{8} = 2$'; $2 \times 2 \times 2 = 8$.

Like square, and its square root, **cube** of a number is NOT always greater than its **cube root.** If a number is $x = 8$, then the '$x^3 = 512$,' and '$\sqrt[3]{x} = 2$'; $512 > 2$; however if a number is $x = 1/8$, then the '$x^3 = 1/512$,' and '$\sqrt[3]{x} = 1/2$;' $1/512 > 1/2$.

Cube root gives unique value of x:

We have seen that if $x^2 = 4, => x = \pm 2.$; no unique value of x; it may be either 2 or -2. However if $x^3 = 8, => x = 2$; a unique value of x; because if $x = -2, x^3 = -8$. So cube root gives the unique value. This concept is widely tested in the Data Sufficiency questions.

n^{th} root:

Like square root, and cube root, n^{th} root means multiplying a number with itself n times to get the original number. Say a number is x, then its nth root would be $\sqrt[n]{x}$.

Example: Say a number is $x = 32$, $\sqrt[5]{32} = 2$; $2 \times 2 \times 2 \times 2 \times 2 = 32$.

Surds

If the root of a number renders an irrational number, its exponent is called **Surd.**

The square root of 2, $\sqrt[2]{2} = 1.414$; an irrational number, hence a surd.

You must know a few values: $\sqrt{2} = 1.414$, $\sqrt{3} = 1.732$, $\sqrt{5} = 2.236$.

3.3 Multiplication and Division of decimals with 10^x

We are often faced with multiplications such as 20.0987×10^3 or $\frac{5,482.75}{10^{-4}}$. Follow the following approach to do such operations in less time.

Scenario 1: Multiplication

Here, we either multiply a number by a number having a positive exponent of 10 or divide it with a negative exponent of 10.

Step 1: Move the decimal to its right for the same number of times as the exponent value.

Step 2: If number of digits to the right of the decimal is less than the value of the exponent, add requisite number of '0s' after the last digit of the number.

Example 1: 20.0987×10^3

The value of the exponent is 3. By moving the decimal to the right 3 times in the number 20.0987, we will get the number 20,098.7.

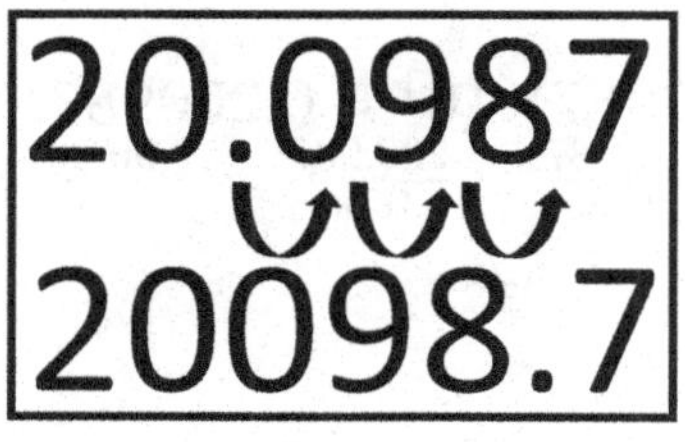

Example 2: 20.09×10^4

As per the process described in the example above, we move the decimal to the right 4 times in the number 20.09. Since the number has only two digits after the decimal, we will add two 0s after '9'. Thus, $20.09 \times 10^4 = 200,900$.

Example 3: $20.0987 \div 10^{-3}$

The value of the exponent is -3. Let us first make it positive. We can bring 10^{-3} in the numerator, thus, $20.0987 \div 10^{-3} = 20.0987 \times 10^3$. Now the situation is the same as was in the example 1. By moving the decimal to the right 3 times in the number 20.0987, we will get the number 20,098.7.

Scenario 2: Division

Here, we either divide a number by a number having a positive exponent of 10 or multiply with a negative exponent of 10.

Step 1: Move decimal to its left for the same number of times as the exponent value

Step 2: If number of digits to the left of the decimal is less than the value of the exponent, add requisite number of '0s' before the first digit of the number

Example 3: 20.0987×10^{-3}

The value of the exponent is -3. By moving decimal to the left 3 times in the number 20.0987, we will get number the 0.0200987.

Example 4: $20.0987 \div 10^{-3}$?

First, convert division into multiplication. So, $20.0987 \div 10^{-3} = \frac{20.0987}{10^{-3}} = 20.0987 \times 10^3$. Now the question fits into scenario 1. Here, the value of the exponent is 3. By moving decimal to the right 3 times in the number 20.0987, we will get the number 20,098.7.

In a nut shell, following numbers are equal; $20.0987 \times 10^3 = 2.00987 \times 10^4 = 0.200987 \times 10^5 = 200.987 \times 10^2 = 20,098.7 = 20,0987 \times 10^{-1}$. Similarly, $\frac{20.0987}{100} = \frac{200.987}{1000} = \frac{2,009.87}{10^4}$.

Example 5: $3,000,000 \times 0.0003$?

To transform the second number, 0.0003, equals to '3 times the exponent of 10,' it needs an exponent of '4' to the base 10 or the multiplication of 10,000; we can borrow four '0s' from the first number, 3,000,000, so $3,000,000 \times 0.0003$ becomes $300\cancel{0000} \times 3$ equalling 900.

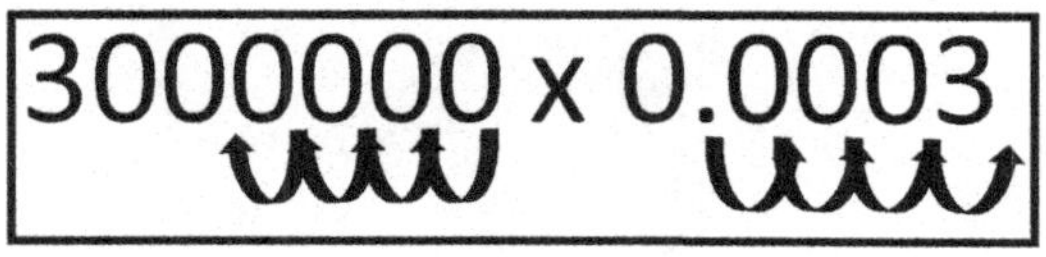

Example 6: $\dfrac{1.96[10^3(6.03 - 3 \times 10^{-2})]}{(1.4)^2} = ?$

$$\Rightarrow \frac{1.96[10^3(6.03 - 3 \times 10^{-2})]}{(1.4)^2} = \frac{\cancel{1.96}[10^3(6.03 - 3 \div 100)]}{\cancel{1.96}} = 10^3(6.03 - 3 \div 100)$$

$\Rightarrow 10^3(6.03 - 0.03) = 1,000 \times 6 = \mathbf{6,000}$.

3.4 Common Binomial expressions

(1) $a(b + c) = ab + ac$

$150 \times 51 = 150(50 + 1) = 150 \times 50 + 150 \times 1 = 7,500 + 150 = 7,650$
$38 \times 19 = 38(20 - 1) = 38 \times 20 - 38 \times 1 = 760 - 38 = 722$

(2) $(a + b)^2 = a^2 + 2ab + b^2$

$102^2 = (100 + 2)^2 = 100^2 + 2 \times 100 \times 2 + 2^2 = 10,000 + 400 + 4 = 10,404$

(3) $(a - b)^2 = a^2 - 2ab + b^2$

$98^2 = (100 - 2)^2 = 100^2 - 2 \times 100 \times 2 + 2^2 = 10,000 - 400 + 4 = 9,604$

(4) $(a^2 - b^2) = (a + b)(a - b)$

$98^2 - 2^2 = (98 + 2)(98 - 2) = 100 \times 96 = 9,600$

(5) $(a + b + c)^2 = a^2 + b^2 + c^2 + 2(ab + bc + ca)$

(6) $(a + b)^3 = a^3 + 3a^2b + 3ab^2 + b^3$

$102^3 = (100 + 2)^3 = 100^3 + 3 \times 100^2 \times 2 + 3 \times 100 \times 2^2 + 2^3 = 1,000,000 + 60,000 + 1,200 + 8 = 1,061,208$

or $(a + b)^3 = a^3 + 3ab(a + b) + b^3$

$51^3 = (50+1)^3 = 50^3 + 3\times 50\times 1\times(50+1) + 1^3 = 125,000 + 150\times 51 + 1 = 132,651$

(7) $(a-b)^3 = a^3 - 3a^2b + 3ab^2 - b^3$

$98^3 = (100-2)^3 = 100^3 - 3\times 100^2\times 2 + 3\times 100\times 2^2 - 2^3 = 1,000,000 - 60,000 + 1,200 - 8 = 941,192$

or $(a-b)^3 = a^3 - 3ab(a-b) - b^3$

$49^3 = (50-1)^3 = 50^3 - 3\times 50\times 1\times(50-1) - 1^3 = 125,000 - 150\times 49 - 1 = 117,649$

(8) $(a^3 + b^3) = (a+b)(a^2 - ab + b^2)$

$11^3 + 9^3 = (11+9)(11^2 - 11\times 9 + 9^2) = 20\times(121 - 99 + 81) = 20\times 103 = 2,060$

(9) $(a^3 - b^3) = (a-b)(a^2 + ab + b^2)$

$11^3 - 9^3 = (11-9)(11^2 + 11\times 9 + 9^2) = 2\times(121 + 99 + 81) = 2\times 301 = 602$

3.5 Fractions

Fraction is a number which is formed out of numerator and denominator. It is represented as $\frac{\text{Numerator}}{\text{Denominator}}$. Both numerator and denominator are integers, but the denominator should not be '0'.
Example: $\frac{3}{4}, \frac{7}{8}, -\frac{9}{5}$, etc.

Proper fraction: For a fraction, if the absolute value of the numerator is less than the absolute value of the denominator, it is called a proper fraction.
Example: $\frac{3}{4}, \frac{7}{8}, -\frac{9}{10}$, etc.

Improper fraction: If the absolute value of the numerator of a fraction is greater than the absolute value of the denominator, it is called improper fraction.
Example: $\frac{4}{3}, \frac{11}{8}, -\frac{9}{5}$, etc..

Mixed fraction: If the fraction is a mix of both an integer and proper fraction, it is called mixed fraction. **Example:** $2\frac{3}{4}, -5\frac{2}{3}$, etc.

Converting mixed fraction to improper fraction: Say, a mixed fraction is $5\frac{2}{3}$.

Step 1: Multiply the integer part with the denominator. $5\times 3 = 15$.

Step 2: Add the numerator to the number attained in step 1. So, $15 + 2 = 17$.
It is the numerator of improper fraction.

Step 3: The equivalent improper fraction is $\frac{\text{new numerator}}{\text{denominator}} = \frac{5 \times 3 + 2}{3} = \frac{17}{3}$

Equivalent Fraction: Fractions that are equal in values are called equivalent fractions, although the numerator and denominator of the equivalent fractions may not be the same.

Example: 1/2 = 2/4 = 3/6 = 4/8 = 5/10.

3.5.1 Concept of fraction

Let us understand the concept of fraction through graphical representation. Fraction is a part of a whole. For example, the fraction 3/8 means that out of 8 parts, only 3 parts are taken; it can also be thought of as eating 3 slices of pizza of an 8-slice pizza.

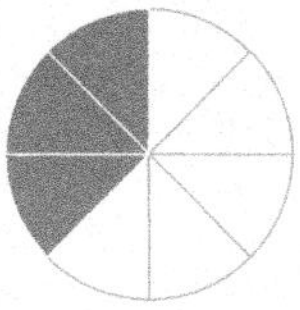

3/8

Reciprocal of fraction: You get reciprocal of a fraction when you interchange numerator and denominator.
Example: Reciprocal of $\frac{3}{8}$ is $\frac{8}{3}$.

What is the significance of $\frac{8}{3}$?

To understand this, first convert $\frac{8}{3}$ into mixed fraction, which is $2\frac{2}{3}$. It means eating 2 full pizzas and 2 slices of pizza of a 3-slice pizza.

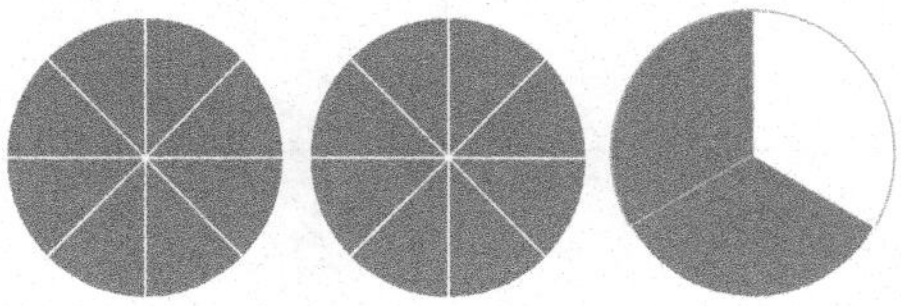

8/3 = Reciprocal of 3/8

Decimal Representation:

A fraction can be represented as a decimal number by dividing the numerator by the denominator. For example, the fraction 3/8 represents the numerical value 0.375; it is a decimal number.

Percent Representation:

A fraction can be represented as a percent by dividing the numerator by the denominator and then multiplying it by 100. For example, $3/4 = 0.75 => 0.75 \times 100\% = 75\%$ and similarly, $1/2 = 50\%$.

3.5.2 Mathematical operations on fractions

Mathematical operation	Example
Addition/	$\frac{1}{2} + \frac{1}{2} = 1$; two halves makes 1;
Subtraction	$\frac{1}{2} - \frac{1}{4} = ? => \frac{1\times2}{2\times2} - \frac{1}{4} = \frac{2}{4} - \frac{1}{4} = \frac{2-1}{4} = \frac{1}{4}$; make denominators equal
Multiplication	$\frac{1}{2} \times \frac{1}{2} = \frac{1}{4}$; half of a half makes it $\frac{1}{4}$; $\frac{3}{5} \times \frac{15}{6} = \frac{3}{2}$; any numerator can cancel any denominator; two numerators will be multiplied, and similarly, two denominators will be multiplied
Division	$\frac{1}{2} \div \frac{1}{2} = \frac{1}{2} \times \frac{2}{1} = 1$; '÷' will change to ' × ' and the fraction to the right will become its reciprocal; $\frac{3}{5} \div \frac{3}{5} = \frac{3}{5} \times \frac{5}{3} = 1$; a fraction divided by itself is always '1'; product of a fraction with its reciprocal is always 1.

Some confusing divisions:

- $\frac{2/3}{3/2} = \frac{2}{3} \div \frac{3}{2} = \frac{2}{3} \times \frac{2}{3} = \frac{4}{9}$;
- $\frac{2}{3/2} = 2 \div \frac{3}{2} = 2 \times \frac{2}{3} = \frac{4}{3}$;
- whereas $\frac{2/3}{3} = \frac{2}{3} \div 3 = \frac{2}{3} \times \frac{1}{3} = \frac{2}{9}$

Example: $\dfrac{2}{2-\dfrac{2}{1-\dfrac{4}{2-\dfrac{5}{2+\frac{6}{7}}}}} = ?$

=> Start solving from the lowermost fraction

$$\Rightarrow \dfrac{2}{2-\dfrac{2}{1-\dfrac{4}{2-\dfrac{5}{2+\frac{6}{7}}}}} = \dfrac{2}{2-\dfrac{2}{1-\dfrac{4}{2-\dfrac{5}{\dfrac{\mathbf{2\times 7+6}}{7}}}}} = \dfrac{2}{2-\dfrac{2}{1-\dfrac{4}{2-\dfrac{5}{\dfrac{\mathbf{20}}{7}}}}};$$

Remember that the primary division operator is between "5" and "20."

So $\frac{5}{\frac{20}{7}} = 5 \div \frac{20}{7} = 5 \times \frac{2}{3} = \frac{7}{20} = \frac{7}{4}$; Do not get it confused with $\frac{\frac{5}{20}}{7} \neq \frac{5}{\frac{20}{7}}$.

Coming back to the question:

$$\Rightarrow \dfrac{2}{2-\dfrac{2}{1-\dfrac{4}{2-\dfrac{5}{\dfrac{\mathbf{20}}{7}}}}} = \dfrac{2}{2-\dfrac{2}{1-\dfrac{4}{2-\dfrac{7\times 5}{20}}}} = \dfrac{2}{2-\dfrac{2}{1-\dfrac{4}{\frac{1}{4}}}} = \dfrac{2}{2-\dfrac{2}{1-4\times 4}} = \dfrac{2}{2-\dfrac{2}{1-16}} =$$

$$\dfrac{2}{2-\dfrac{2}{-15}}$$

$$\Rightarrow \dfrac{2}{2+\dfrac{2}{15}} = \dfrac{2}{\dfrac{32}{15}} = \dfrac{2\times 15}{30} = \dfrac{15}{16}$$

3.5.3 Comparing fractions

You may come across situation where you have to compare the two fractions—which fraction is greater or smaller?

A general rule: For positive factions, larger the numerator, greater is the fraction (with the same denominator), and smaller the denominator, greater is the fraction (with the same numerator) and vice-versa; opposite is true for negative fractions.

Example 1: Which is greater $\frac{1}{2}$ or $\frac{1}{3}$? Well, this is a simple one as the numerators of the fractions are equal. For positive factions, if the numerators of the fractions are equal, the fraction with smaller denominator is greater, so $\frac{1}{2} > \frac{1}{3}$; however $-\frac{1}{2} < \frac{1}{3}; -\frac{1}{2} <$

$-\frac{1}{3}; \frac{1}{2} > -\frac{1}{3}$.

Example 2: Which is greater $\frac{2}{5}$ or $\frac{3}{5}$? Again, this is a simple one as the denominator of the fractions are equal. For positive factions, if the denominators of the fractions are equal, the fraction with the larger numerator is greater, $\frac{2}{5} < \frac{3}{5}$; however $\frac{2}{5} > -\frac{3}{5}; -\frac{2}{5} > -\frac{3}{5}; \frac{2}{5} > -\frac{3}{5}$.

Example 3: Which is greater $\frac{11}{15}$ or $\frac{9}{16}$? This is the easiest of all to deduce. As the numerator of the first fraction is greater than that of the other and its denominator is smaller than that of the other, so on both the counts, the first fraction scores over the other, hence it is greater than the second fraction, so $\frac{11}{15} > \frac{9}{16}$; however $-\frac{11}{15} < -\frac{9}{16}$.

Example 4: Which is greater $\frac{5}{7}$ or $\frac{7}{9}$? This is not an easier one to deduce. The numerator of the first fraction is greater than that of the other but its denominator is also larger than that of the other, so on one count, the first fraction scores over the second and on the other count, the second one scores over the first. So how to deduce which fraction is greater? There are three approaches to deduce this.

Approach 1:

Convert fractions to Decimals

Each fraction can be converted into a decimal by dividing the numerator by the denominator. In decimal format, the value of each fraction can be compared.

$\Rightarrow \frac{5}{7} = 0.714; \frac{7}{9} = 0.777$, so $0.777 > 0.714$.

Since the decimal equivalent of 7/9 is larger than that of 5/7, so 7/9 > 5/7.

Approach 2:

Take 50%

The approach is to take 50% of the denominator and deduct it from the numerator, and then compare the remaining fractions.

Which is greater $\frac{5}{7}$ or $\frac{4}{5}$?

=> 50% of 7 (denominator of the first fraction) is 3.5, and 50% of 5 (denominator of the second fraction) is 2.5, so we can write $\left[\frac{5}{7} = \frac{3.5}{7} + \frac{1.5}{7}\right] <> \left[\frac{4}{5} = \frac{2.5}{5} + \frac{1.5}{5}\right]$

=> $(50\% + \frac{1.5}{7}) <> (50\% + \frac{1.5}{5})$ => It is obvious that $\frac{1.5}{7} < \frac{1.5}{5}$, so, $\frac{5}{7} < \frac{4}{5}$.

This approach may not be applicable in each case; after seeing all the approaches, you should decide which approach works best for a specific scenario.

Approach 3:

Cross Multiplication

Another approach to compare fractions is: Cross Multiplication. Numerator of the first fraction is multiplied to the denominator of the second fraction, and numerator of the second fraction is multiplied to the denominator of the first fraction. Now compare the results, and identify the greater of the two products; the fraction which has its numerator in the greater of the two products is a greater fraction than the other fraction.

=> Which is greater $\frac{5}{7}$ or $\frac{7}{9}$?

=> (Numerator of I × Denominator of II) = $5 \times 9 = 45$; (Numerator of II × Denominator of I) = $7 \times 7 = 49$, since $45 < 49$ and "49" includes the numerator of the second fraction, hence the second fraction is greater than the first fraction.

3.5.4 More on fractions

(1) If $\frac{a}{b} > 1$, then $\frac{a+c}{b+c} < \frac{a}{b}$; where $a, b, \&\ c$ are positive numbers

=> Say $\frac{a}{b} = \frac{3}{2}$, and $c = 1$, then $\frac{a+c}{b+c} = \frac{3+1}{2+1} = \frac{4}{3} = 1.33 < \left[\frac{a}{b} = \frac{3}{2} = 1.5\right]$

We can understand it this way: Since the fraction $a/b > 1$, it means that $a > b$. When we add c to each a and b, larger proportion of c is added to b i.e. denominator and relatively a smaller proportion of c is added to a i.e. the numerator. Let us see how.

$$\frac{a+c}{b+c} = \frac{a\left(1+\frac{c}{a}\right)}{b\left(1+\frac{c}{b}\right)} = \frac{a}{b} \times \frac{\left(1 + \text{smaller fraction compared to } \frac{c}{b}\right)}{\left(1 + \text{larger fraction compared to } \frac{c}{a}\right)}$$

$$= \left[\frac{a}{b} \times \frac{\text{smaller number}}{\text{larger number}}\right] < \frac{a}{b}$$

(2) If $0 < \frac{a}{b} < 1$, then $\frac{a+c}{b+c} > \frac{a}{b}$; where $a, b, \&\ c$ are positive numbers

=> Say $\frac{a}{b} = \frac{2}{3}$, and $c = 1$, then $\frac{a+c}{b+c} = \frac{2+1}{3+1} = \frac{3}{4} = 0.75 > \left[\frac{a}{b} = \frac{2}{3} = 0.667\right]$

(3) If $\frac{a}{b} > 1$, then $\frac{a-c}{b-c} > \frac{a}{b}$; where $a, b, \&\ c$ are positive numbers

=> Say $\frac{a}{b} = \frac{3}{2}$, and $c = 1$, then $\frac{a-c}{b-c} = \frac{3-1}{2-1} = \frac{2}{1} = 2 > \left[\frac{a}{b} = \frac{3}{2} = 1.50\right]$

(4) If $0 < \frac{a}{b} < 1$, then $\frac{a-c}{b-c} < \frac{a}{b}$; where $a, b, \& \ c$ are positive numbers

=> Say $\frac{a}{b} = \frac{2}{3}$, and $c = 1$, then $\frac{a-c}{b-c} = \frac{2-1}{3-1} = \frac{1}{2} = 0.50 < \left[\frac{a}{b} = \frac{2}{3} = 0.667\right]$

3.5.5 Fractions and their percent equivalents

Fraction	Percent
$\frac{1}{1}$	100%
$\frac{1}{2}$	50%
$\frac{1}{3}$	33.33%
$\frac{1}{4}$	25%
$\frac{1}{5}$	20%
$\frac{1}{6}$	16.67%
$\frac{1}{7}$	14.28%
$\frac{1}{8}$	12.50%

Fraction	Percent
$\frac{1}{9}$	11.11%
$\frac{1}{10}$	10%
$\frac{1}{11}$	9.09%
$\frac{1}{12}$	8.33%
$\frac{1}{25}$	4%
$\frac{1}{50}$	2%
$\frac{1}{100}$	1%
$\frac{1}{200}$	0.50%

Application: Sometimes you may come across calculating an odd-looking percent of an odd-looking number. The fraction to percent equivalent table may help you make the calculations easier.

Example 1: What is 14.25% of 705?

The traditional way of calculation is time-consuming: $\frac{14.25}{100} \times 705 = 100.47$. Let us apply a smarter approach.

Smarter approach:

From the table, we know that 14.25% is $\approx$ 14.28% which is equal to $\frac{1}{7}$. So, we can rather calculate $\frac{1}{7}$ of 705; again, we can approximate 705 to 707 as there is a difference of only '2' between 705 and 707. So, $\frac{1}{7}$ of 707 is easier to calculate which is equal to '101'. But '101' is an approximate value, the actual value would be less than '101' as we increased 14.25% to 14.28% and 705 to 707. The actual value already calculated as 100.47, which is quite close to the approximate value '101'.

Example 2: What is 12.61% of 802?

A. 95.68

B. 105.94

C. 101.13

Again, from the table, we know that 12.61% is $\approx$ 12.50%, which is $\frac{1}{8}$. So, we can rather calculate $\frac{1}{8}$ of 802; again, we can approximate 802 as 800 as there is a difference of only '2' between them. So, 1/8 of 800 is easier to calculate which is equal to '100'. Again, '100' is an approximate value, the actual value must be little more than '100' as we had decreased 12.61% to 12.50% and 802 to 800. Looking at the options, we can safely lock '101.13' as the actual value since the other higher value is far greater than '100'.

3.5.6 See-Saw Approach

How do you solve a question, which asks, "*If gasoline price increased by 10%, by what percent should one reduce its consumption so that the expense on gasoline remains the same?*"

Traditional Approach:

Say, before the increase, the price of gasoline was \$100/gallon, and the consumption was 100 gallons. So, the expense = $100 \times 100 = \$10,000$.

Now, after the increase, the price of gasoline is \$110/gallon (10% up), and say the consumption is reduced to x gallons. So, the expense = $110 \times x = \$110x$.

Since the expense needs to be the same as before, hence $10,000 = 110x$ => $x = 90.90$ gallons

=> Percent reduction in consumption = $\dfrac{100 - 90.90}{100} \times 100\% = 9.09\%$.

See-Saw Approach:

Scenario 1: When $x \times y = C$; where x is one element, y is second element, and C is a constant.

If x goes up by certain percent and y must go down with some percent such that after the change $x \times y = x' \times y' = C$, follow the following short-cut steps to get the answer. Where after the change, x becomes x', and y becomes y'.

Step 1: Covert the given percent figure into fraction, say it is $\frac{1}{a}$; you should take the help of the conversion table.
Step 2: Add "1" to the denominator of the fraction attained; it would be $\frac{1}{a+1}$.
Step 3: Convert $\frac{1}{a+1}$ fraction into percent; it is the requisite answer; you should take the help of the table.

Let us get the answer to the above question with the See-Saw approach.

Step 1: Price goes up by 10%; equivalent fraction of 10% is $\frac{1}{10}$, so $\frac{1}{a} = \frac{1}{10}$.
Step 2: By adding "1" to the denominator of the fraction $\frac{1}{a}$, we get $\frac{1}{a+1} = \frac{1}{11}$.
Step 3: With the help of the table, we get $\frac{1}{11} = 9.09\%$; it is the requisite answer. The consumption must go down by 9.09%.

Pl. see the pictorial representation of the See-Saw approach.

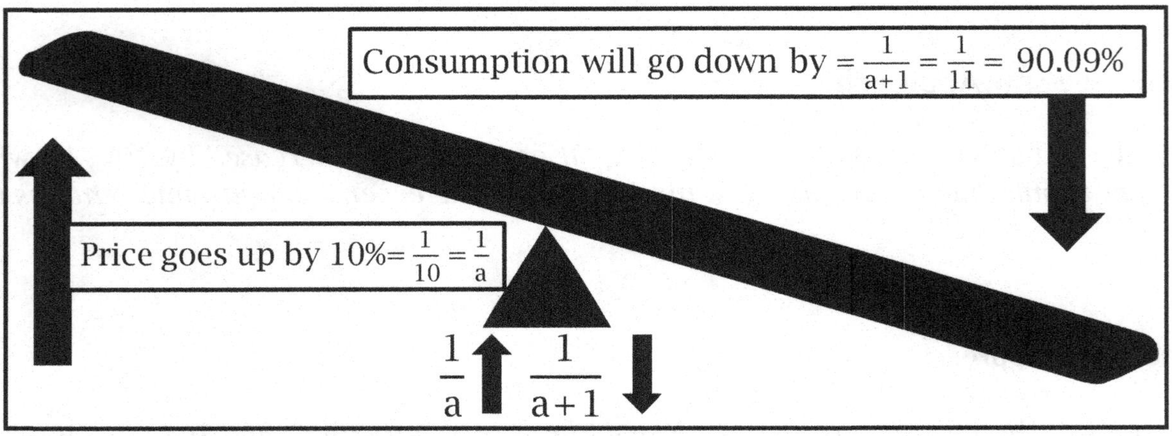

Scenario 2: When $x \times y = C$; where x is one element, y is second element, and C is constant.

If x goes down by certain percent and y must go up with some percent such that after the change $x \times y = x' \times y' = C$, follow the following short-cut steps to get the answer.

Step 1: Covert the given percent figure into fraction, say it is $\frac{1}{a}$; take the help of the table.
Step 2: Subtract "1" from the denominator of the fraction attained; it would be $\frac{1}{a-1}$

Step 3: Convert $\dfrac{1}{a-1}$ fraction into percent; it is the requisite answer; you should take the help of the table.

Let us apply the See-Saw approach in another question: "*If milk price decreased by 10%, by what percent should one increase its consumption so that the expense on milk remains the same?*"

Step 1: Price goes down by 10%; equivalent fraction of 10% is $\frac{1}{10}$, $\frac{1}{10}$, so $\frac{1}{a} = \frac{1}{10}$;
Step 2: By subtracting "1" from the denominator of the fraction $\frac{1}{a}$, we get $\dfrac{1}{a-1} = \frac{1}{9}$.
Step 3: With the help of the table, we get $\frac{1}{9} = 11.11\%$; it is the requisite answer. The consumption must go up by 11.11%.

See the pictorial representation of the same.

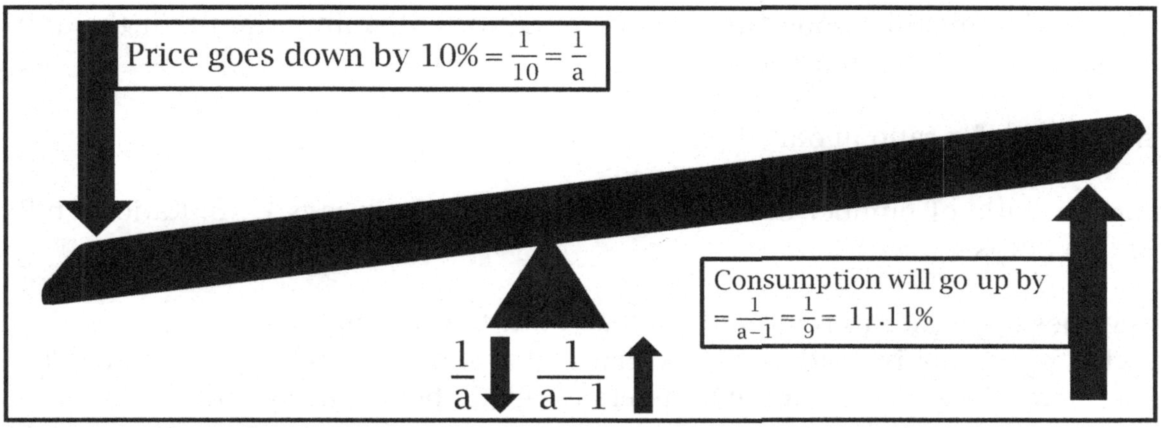

Example 1: If the sale of a brand is reduced by 17%, by what percent should the company increase its sales price such that the revenue earned remains the same as before?

A. 14.28%

B. 14.53%

C. 20.00%

D. 20.48%

First, identify whether the question fits into the category discussed above. Since "Quantity × Price = Revenue," and the condition is that the revenue must remain constant, hence the question fits into the See-Saw approach model. Out of the two scenarios, the second scenario fits as an element goes down and the complementary element must go up.

When we look at the table, we do not find 17%, however we can safely assume that it is closer to 16.67%; so the equivalent of 17% $=\approx$ 16.67% $=\approx \frac{1}{6} = \frac{1}{a} => \frac{1}{a-1} = \frac{1}{5} = 20\%$. The sales price must be increased by 20%. The correct answer would be a little more than 20% as we approximated 17% to 16.67%. So, the correct answer must be option D—the only greater option than 20%. The company should increase the sales price of the brand by 20.48%.

Example 2: If the price of milk increased by 11%, by what percent should its consumption be reduced such that the expenses on milk remains the same as before?

We identified that the question fits into the category discussed above, and it fits into the first scenario. Now, 11% $=\approx$ 11.11% $= \frac{1}{9} = \frac{1}{a} => \frac{1}{a+1} = \frac{1}{10} =\approx$ 10%.. The correct answer would be a little less than 10%.

3.6 Ratio & Proportion

Ratio is a relationship among two or more quantities; it is important that the units of the quantities considered are same.

We normally refer ratio in daily life.

Examples: Ratio of number of boys to that of girls in a class is 2 : 3; Ratio of milk to water in a pot is 4 : 7, etc.

Ratio values are easier to refer and compare, but they are not absolute values. For the sake of comparison, instead of saying that region X has 100,000 cars and 50,000 two-wheelers; we can reduce both the numbers to relate them conveniently.

$$\frac{\text{\# of Cars}}{\text{\# of Two-wheelers}} = \frac{C}{T} = \frac{100,000}{50,000} = \frac{2}{1}$$

=> C : T = 2 : 1; It means that for every two-wheeler, there are two cars in the region X.

Terms used:

The first term, 2, of the ratio C : T = 2 : 1 is called **antecedent**, and the second term, 1, is called **consequent.**

Example 1: Two quantities are in the ratio of 5 : 7. If the first quantity is 155, find the second quantity.

Ratio values				Actual values		
Qty 1	:	Qty 2	::	Qty 1	:	Qty 2
5	:	7	::	155	:	? (x)

=> $\frac{5}{7} = \frac{155}{x}$

=> $5x = 7 \times 155$ => $x = \frac{7 \times 155}{5} = 217$

Some rules:

(1) Multiplying the antecedent and the consequent of the ratio with the same number does not make it a different ratio; it will be called an equivalent ratio.

$$\Rightarrow \frac{C}{T} = \frac{2}{1} = \frac{2 \times 10}{1 \times 10} = \frac{20}{10} = \frac{30}{15} = \frac{40}{20} = \frac{100,000}{50,000}$$

(2) We cannot get the actual values of the quantities compared from the ratio; we can only compare their relative values.

=> From the ratio C : T = 2 : 1, we cannot assume that the actual value of number of cars is 100,000 and that of two-wheelers is 50,000.

(3) Units of the quantities compared must be the same.

=> If the weights of three objects A, B and C are 1 ton, 500 kilograms, and 500 grams respectively, then what is the ratio of their weights?

First, convert each unit of measurement into single units of measurement convenient to you. It's better that we convert ton and gram into kilogram.

=> Weight of object A = 1 ton = 1,000 kg.; Weight of object C = 500 grams = 1/2 kg.

So, the ratio of Wt. of A : Wt. of B : Wt. of C :: 1,000 : 500: 1/2 => 2,000 : 1,000 : 1 (Multiplying by "2" throughout to get integer values.)

Example 2: A cask contains an 80 liter mixture of milk and water in the ratio of 69 : 31. By what percent of the mixture is milk more than water?

=> Given is, M : W :: 69 : 31; we want $\frac{\textbf{Milk - Water}}{\textbf{Mixture (80)}} \times 100\%$. It would be time-consuming to calculate the quantity of milk and water as the ratio values are in odd numbers. Let us apply the following approach.

=> **Ratio scale:**

M	:	W	:	Mixture (M + W)	:	(M - W)
69	:	31	:	(69 + 31)	:	(69 - 31)
				100	:	38

=> answer is (38/100) × 100% = 38%.

3.6.1 Proportion

Proportion is the equality of two ratios. **Example:** 2 : 3 :: 4 : 6.

Types of proportions:

(A) **Direct proportion:** Two quantities are said to be in direct proportion if an increase in one quantity results in a proportional increase in the other quantity; and similarly, a decrease in one quantity results in a proportional decrease in the other quantity.

Example: If the price of one phone is \$300, the cost of two phones would be \$600.

Phone1 : Phone2 :: Price1 : Price2

1 : 2 :: 300 : ?(x)

$$=> \frac{1}{2} = \frac{300}{x} => x = 2 \times 300 = \$600.$$

(B) **Inverse proportion:** Two quantities are said to be in indirect or inverse proportion if an increase in one quantity results in a proportional decrease in the other quantity, and vice versa.

Example 3: If 3 men can complete a task in 2 days, in how many days will 6 men complete the task?

Men1 : Men2 :: Days2 : Days1

3 : 6 :: ?(x) : 2

$$=> \frac{3}{6} = \frac{x}{2} => x = 1 \text{ day.}$$

(C) **Variation:** A quantity in ratio may increase or decrease w.r.t. power or exponent of other quantity.

Example 4: If the price of diamond varies with square of the size of the diamond, and the price of a diamond A is \$1,000, then what is the price of the diamond B which is double the size of diamond A?

$$=> \text{Price} \propto (\text{Size})^2 => \text{Price} = K \times (\text{Size})^2; \text{ where K is a constant}$$

So, we can write, PriceA $= K \times (\text{SizeA})^2$ & PriceB $= K \times (\text{SizeB})^2$

So, $\dfrac{\text{PriceA}}{\text{PriceB}} = \dfrac{K \times (\text{SizeA})^2}{K \times (\text{SizeB})^2}$

PriceA	:	PriceB	::	$(\text{SizeA})^2$	:	$(\text{SizeB})^2$
1,000	:	?(x)	::	1^2	:	2^2
1,000	:	?(x)	::	1	:	4

=> $x = 4 \times 1,000 = \$4,000$; Assume that the size of diamond A is 1 unit.

Diamond B will cost $4,000.

Example 5: If the efficiency of a machine varies inversely with the cube of length of time it worked, and its efficiency is 90% for the 6 hours it worked, then what would be its efficiency if it worked for 10 hours?

=> Efficiency $\propto \dfrac{1}{(\text{time})^3}$ => Eff. $= \dfrac{K}{(\text{time})^3}$; where K is a constant

So, we can write, Eff.(90%) $= \dfrac{K}{(\text{time@90\%})^3}$, and Eff.($x$%) $= \dfrac{K}{(\text{time@}x\%)^3}$

So, $\dfrac{\text{Eff.(90\%)}}{\text{Eff.}(x\%)} = \dfrac{(\text{time@}x\%)^3}{(\text{time@90\%})^3}$

Eff.(90%)	:	Eff.(x%)	::	$(\text{time@}x\%)^3$	:	$(\text{time@90\%})^3$
90%	:	?(x)	::	10^3	:	6^3

=> $x = \dfrac{90\% \times 6 \times 6 \times 6}{10 \times 10 \times 10} = 19.44\%$

3.6.2 Role of common factor

Example 6: Two quantities are in the ratio of 5 : 7. If the first quantity is 155, find the second quantity.

Though we have dealt this question before, for the sake of understanding the role of common factor, we look at it again. Since the antecedent, 5, and the consequent, 7, show relative values, hence there must be a common factor between them; say it is x.

So, the actual values would be $5x$, and $7x$, hence we can deduce that $5x = 155 => x = 31$; so $7x = 7 \times 31 = 217$. The second quantity is 217.

Example 7: Two quantities are in the ratio of 2 : 7. If each of them is increased by 14, then their ratio become 4 : 7. Find the numbers.

Say the common factor is x, so the quantities are $2x$ and $7x$. After the increase, the quantities become $2x + 14$ and $7x + 14$. So we can write:

$$=> \frac{2x + 14}{7x + 14} = \frac{4}{7} => 7(2x + 14) = 4(7x + 14) => x = 3.$$

So the quantities are $2x = 2 \times 3 = 6$ and $7x = 7 \times 3 = 21$.

3.7 Percents

Percent is the most used concept in Arithmetic. When we quote a relative figure, we mostly refer it w.r.t. 100.

Example: In a class, there are 20 boys and 30 girls. What is the percent of boys in the class?

$$\% = \frac{\text{Value}}{\text{Total}} \times 100\% = \frac{\text{Number of boys}}{\text{Total number of students}} \times 100\%$$

$$= \frac{20}{20 + 30} \times 100\% = \frac{20}{50} \times 100\% = 40\%.$$

3.7.1 Smart calculation

Calculating the percent values may be time-consuming for some. Here are few smart techniques to make things easier. If you wish to calculate 10% of 5,640, you may do it traditionally: 10% of $5,640 = \frac{10}{100} \times 5,640 = 564$.

The table below shows how to calculate some common percent values with a smarter approach.

Calculating x% of a number	**Smarter approach**
10% of a number	Move decimal one place left. So, 10% of 5,640 would be 564.
5% of a number	Calculate 10% and then make it half. So, 5% of 5,640 would be (10% of 5,640)/2 = 564/2 = 282.
1% of a number	Move decimal two places left. So, 1% of 5,640 would be 56.40.
2% of a number	Calculate 1% and then double it. So, 2% of 5,640 would be 2× (1% of 5,640) = 2×56.40 = 112.80.
0.5% of a number	Calculate 1% and divide it by 2. So, 0.5% of 5,640 would be 56.40/2 = 28.20.
11% of a number	Calculate 10% and 1%, and then add them. So, 11% of 5,640 would be (10% of 5,640 + 1% of 5,640) = 564 + 56.40 = 620.40.
100% of a number	100% of the number is number itself. So, 100% of 5,640 would be 5,640.
50% of a number	Calculate one-half of the number. So, 50% of 5,640 would be (1/2 of 5,640) = 2,820.

25% of a number	Calculate one-fourth of the number. So, 25% of 5,640 would be (1/4 of 5,640) = 1,410.
12.50% of a number	Calculate 25% and then make it half. So, 12.5% of 5,640 would be (25% of 5,640)/2 = 1,410/2 = 705. Alternatively, you can calculate one-eight of the number. So, 12.5% of 5,640 would be (1/8 of 5,640) = 705.
33.33% of a number	Calculate one-third of the number. So, 33.33% of 5,640 would be (1/3 of 5,640) = 1,880.
75% of a number	Calculate three-fourth of the number. So, 75% of 5,640 would be (3/4 of 5,640) = 3×1410 = 4,230. Alternatively, you can calculate one-fourth of the number, and deduct it from the number. So, 75% of 5,640 would be (5,640 - 1/4 of 5,640) = 4,230.
20% of a number	Calculate one-fifth of the number. So, 20% of 5,640 would be (1/5 of 5,640) = 1,128.
90% of a number	Calculate 10% of the number and deduct it from the number. So, 90% of 5,640 would be (5,640 - 10% of 5,640) = 5,076.

3.7.2 Some applications

(A) **Percentage increase or decrease**

You may have to calculate the percent increase or decrease in value. Say, for example, if the population of a city rose from 45 million in the year 2011 to 50 million in the year 2012, then what is the percentage increase in the population of the city?

$$\Rightarrow \textbf{\% increase} = \frac{\textbf{New value - Base value}}{\textbf{Base value}} \times 100\%$$

$$\Rightarrow \textbf{\% increase in population} = \frac{\textbf{Popn. in yr. 2012 - Popn. in yr. 2011}}{\textbf{Population in yr. 2011}} \times 100\%$$

$$\Rightarrow \textbf{\% increase in population} = \frac{50-45}{45} \times 100\% = \frac{5}{45} \times 100\% = 11.11\%$$

(B) **Beware of word *than***

If the question asks—*By what percent is the population in the year 2011 less than that in the year 2012?*

Will the answer be 11.11%?—No. It is not. In this question, the base of comparison is changed to the year 2012. Year 2012 will appear as base in the denominator.

$$\Rightarrow \textbf{\% decrease in population} = \frac{\textbf{Popn. in yr. 2012 - Popn. in yr. 2011}}{\textbf{Population in yr. 2012}} \times 100\%$$

$$\Rightarrow \textbf{\% decrease in population} = \frac{50-45}{50} \times 100\% = \frac{5}{50} \times 100\% = 10\%$$

(C) **Not always Percentage increase or decrease**

(a) If the question asks—*What percent is the population in the year 2011 of the population in the year 2012?*

We need NOT calculate % increase or decrease here.

$$\Rightarrow \frac{\textbf{Population in yr. 2011}}{\textbf{Population in yr. 2012}} \times 100\% = \frac{45}{50} \times 100\% = 90\%$$

(b) Similarly, if the question asks — *What percent is the population in the year 2012 of the population in the year 2011?*

$$\Rightarrow \frac{\textbf{Population in yr. 2012}}{\textbf{Population in yr. 2011}} \times 100\% = \frac{50}{45} \times 100\% = 111.11\%$$

(D) **Determine increased or decreased value**

(a) If the question asks — *What would be the population in the year 2013, if the population grows at the same rate as during the period 2011-2012?*

We have already calculated % increase in the population equaling 11.11%. Applying traditional approach to solve this question is time-consuming. As per the traditional approach, we calculate the increase first, and then add the increase to the population in the year 2012 to get the population in the year 2013.

Following approach will save your precious time.

=> $\textbf{Population}|_{2013} = \textbf{Population}|_{2012} \times (1 + \%\ \textbf{increase})$

=> $50 \times (1 + 11.11\%) = 50 \times (1 + 0.111) = 50 \times 1.11 = 55.55$ million

(b) Similarly, if the question is — *What would be company X's revenue for next year if it decreased by 23.87% from the current year revenue of $237.48 million?*

Following approach will save your precious time.

=> $\textbf{Revenue}|_{\text{nxt. Yr.}} = \textbf{Revenue}|_{\text{Current Yr.}} \times (1 - \%\ \textbf{increase})$

=> $\text{Revenue}|_{\text{nxt. Yr.}} = 237.48 \times (1 - 0.2389) = 237.48 \times 0.7613$

Calculating 237.48×0.7613 without the help of calculator is frustrating. You can even save more time by truncating the decimals and approximating the number.

=> $\text{Revenue}|_{\text{nxt. Yr.}}$

$= 237.48 \times (1 - \approx 0.24) = 237.48 \times \approx 0.76 = 237.48 \times 0.75 = \approx 240 \times \frac{3}{4} = \approx 180$ million

(E) **Percent – Percent ≠ Percent**

If the question asks—*A shopkeeper used to charge 15% margin on its products. Due to competition, he revised the margin to 10%. By what percent did the shopkeeper decrease his margin?*

The answer is certainly not 5%.

Say, the sales is \$100, then the margin charged earlier = 15% of 100 = \$15, and the margin charged now = 10% of 100 = \$10.

% decrease in margin = $\frac{15-10}{15} \times 100\% = 33.33\%$. So, the shopkeeper reduced its margin by 33.33%.

So, what does 15% – 10% represent?

In fact, **Percent – Percent = Percent Point**

Alternatively, we can also say that the shopkeeper reduced his margin by **5% points.**

(F) **Percent of what**

If the question asks—*If 80% of students play soccer and 25% of them joined a professional soccer club, what percentage of of the total students in the school did not join the club?*

The answer is not 75% (100% – 25%). We must pay attention to the verbiage. It is not always that an entity is a percentage of total.

Here, the number of students joined club = 25% of 80% = $\frac{25}{100} \times 80\% = 20\%$. So, 20% joined the club, and 80% of students did not join the club.

(G) **Successive percent**

Calculating percents one after another is called successive percents. This concept is applied in case of calculating 'successive discounts' or 'successive profits' and also in cases when 'a quantity is expressed as a product of two other quantities'.

If the question asks—*A shopkeeper offers a discount of 20% on the MRP, and then further a discount of 10% on the invoice value. How much a customer will have to pay for a camera whose MRP is \$200?*

Total discount offered by the shopkeeper is not 30% (20% + 10%). The reference values on which each discount was offered is different; while the first discount was offered on the MRP, the other one was offered on the invoice value.

Say the MRP is \$100; so the invoice value after discount = MRP - Discount = 100 - 20 = \$80, hence amount paid by the customer = Invoice value - II discount = 80 – 10% of 80 = 80 – 8 = 72%. So the customer paid 72% of \$200 = \$144 for the camera.

If there are $d_1, d_2, \& \ d_3$ percentage successive discounts, then the final price can be calculated in the following way.

=> Price after discount = MRP $\left(1 - \frac{\mathbf{d_1}}{\mathbf{100}}\right)\left(1 - \frac{\mathbf{d_2}}{\mathbf{100}}\right)\left(1 - \frac{\mathbf{d_3}}{\mathbf{100}}\right)$

Alternate approach:

If $x\%$ and $y\%$ are two successive percentage changes, the overall percentage change is given by:

$$\left(x + y + \frac{x \times y}{100}\right)\%;$$

where x and y are taken with their sign i.e. '+' for increase and '–' for decrease.

Applying this in the above question:

Total discount $= -20 - 10 + \frac{(-20) \times (-10)}{100} = -28\%$ discount.

Thus, the customer pays $(100 - 28)\%$ of \$200 = 72% of \$200 = \$144.

Another example:

Say, the length of a rectangle is increased by 10% and the breadth is decreased by 5%. What is the percentage change in the area of the rectangle?

We know that Area of Rectangle = Length × Breadth i.e. a quantity (Area) is expressed as a product of two quantities (Length and Breadth). Thus, we can use the concept of successive % change.

The overall percentage change in area $= +10 - 5 + \frac{10 \times (-5)}{100} = 5 - 0.5 = 4.5\%$ (Positive implies an increase in area).

(H) **Number of times and percent: Few short cuts:**

Question: If X is increased by 3 times, X now is what percent of the original value?	
Traditional approach	**Smarter approach**
Say, X = 100, then $X_{\text{now}} = 100 + 3 \times 100 = 400$; $X_{\text{now}} = \frac{400}{100} \times 100\%$ of X = 400% of X	If X increased by n times, then $X_{\text{now}} = (1 + n) \times 100\%$ of X; Or, X increased by $100n\%$. $X_{\text{now}} = (1 + 3) \times 100\%$ of X = 400% of X
Question: If X is increased by 300%, how many times is X now of the original value?	
Traditional approach	**Smarter approach**
Say, X =100, then $X_{\text{now}} = 100 + 300\%$ of $100 = 400$; $X_{\text{now}} = \frac{400}{100} \times X = 4X$	If X increased by y %, then $X_{\text{now}} = (1 + \frac{y}{100}) \times X$; Or, X increased by $\frac{y}{100}$ times. $X_{\text{now}} = (1 + 3) \times X = 4X$
Question: Price of a commodity costing \$250 increased by 10%. What is the price after increase?	
Traditional approach	**Smarter approach**
New Price = Old Price + Increase; $= 250 + \left(250 \times \frac{10}{100}\right) = \$275$	If the price is increased by y%, then New Price = Old Price $\times$ (1 + y%) $= 250 \times 1.1 = \$275$; Or, X increased by y%. New Price = 110% of Old Price

A few examples:

(i) X is less than Y by 30% => It is same as saying X = 70% of Y => X = 0.7Y.

(ii) X is more than Y by 230% => It is better that you do as X = 330% of Y => X = 3.3Y.

(iii) X is less than Y by 60%, and Z is more than X by 50%.

It is better that you do as X = 40% of Y => X = 0.4Y, and Z = 150% of X

=> Z = 1.5X => Z = 1.5 × 0.4Y = 0.6Y.

3.8 Translating words into math

Often times, it is observed that once you have an algebraic equation in hand, it is easier to solve it and get the desired value(s) of the variable(s). However, especially in arithmetic, you will find that the math is trapped in the English language, and you have to extract math out of it. We bring you some examples to help you understand the math part of the question narrations.

Example 1: What is 75% of 200?

The question wants us to find out the value of 75% of the number 200. Let us interpret each word in the question narration. “What = ?”, “is = =”, & “of = × (Multiplication)”

What	is	75 %	of	200	?
? Say x	=	75%	×	200	$x = 75\% \times 200$

Example 2: 150 is how much percent of 200?

150	is	how much %	of	200	?
150	=	Say x%	×	200	$150 = x\% \times 200$

Example 3: Find a fraction that bears the same ratio to 1/27 that 3/7 does to 5/9.

Find a fraction that bears the same ratio	to	1/27	that	3/7	does to	5/9	.
Say x	:	1/27	=	3/7	:	5/9	$\frac{\frac{x}{1}}{27} = \frac{\frac{3}{7}}{\frac{5}{9}}$ $\Rightarrow 27x = \frac{3 \times 9}{7 \times 5}$

Example 4: If A earns 30% more than B, by what percent is B's income less than that of A?

A	earns	30%	more than	B,	by what percent	is	B's income	less than that of A?
A	=	30%	+	B,	x	=	(B	$\frac{(A-B)}{A} \times 100\%$
A = 130				Say B = 100	$x = \left(\frac{A-B}{A}\right) \times 100\%$; Taking care of sign			

Example 5: Fee for an exhibition was $2. It was reduced by 25%, resulting into an increase in sales revenue by 20%. Find the percentage increase in the number of visitors.

Fee for an exhibition	was	$2.	It was reduced by 25%,	this resulted in increase in sales revenue by 20%.	Find the percentage increase in the number of visitors.
x	=	$2	$y = x - 25\%$ of x $= 75\%$ of x $= 0.75x$ $= 0.75 \times 2$ $= 1.50$	Say sales (before) = $100; Sales (now) = $120	$\frac{\#\text{(now)} - \#\text{(before)}}{\#\text{(before)}} \times 100\%$ #(before) = $100/2 = 50; #(now) = $120/1.50 = 80

3.9 Interest

Key terms:

Principal or Sum borrowed: The money borrowed is called Principal or Sum. It is normally denoted by a symbol P.

Interest: The money paid back over and above the borrowed sum is called Interest. It is normally denoted by a symbol I.

Amount: The money paid back to the lender along with the interest is called Amount. Amount = Sum + Interest. It is normally denoted by a symbol A.

Rate of interest:

'100' is a base or reference value for calculating interest. So if the principal = 100, and the interest = 10, then we can say that the rate of interest is 10%. It is normally denoted by a symbol r.

Period of borrowing:

The money borrowed is usually taken on loan for a certain period, say-one year, 6 months, 3 months, 1 month, or even for a few days. So it is important to specify that for what period of time, the rate of interest is applicable. Saying 'r = 10%' is an incomplete information; one must say 'r = 10% per annum' or some specific time period. But it does mean that the period of borrowing is also one year; 'r = 10% per annum' does not have relation to the period of borrowing.

3.9.1 Simple Interest

Say a sum of $1,000 is borrowed at a rate of 10% per annum for 2 years; the interest for the first year would be $\left[1,000 \times \frac{10}{100} = \$100\right]$. Similarly the interest for the second year would be another $100. So total interest for 2 years = $200. We can make a formula out of it.

$$SI = \frac{P \times r \times t}{100}$$

where SI is simple interest, P is sum borrowed, and t is time period.

Example 1: A man borrowed $16,000 at a rate of $12\frac{1}{2}$% per annum for $6\frac{1}{2}$ years. What amount shall he pay back to the lender?

$$\Rightarrow SI = \frac{P \times r \times t}{100} = \frac{16,000 \times (25/2) \times (13/2)}{100} = \frac{16,000 \times 25 \times 13}{100 \times 2 \times 2} = \$13,000$$

=> Amount A = Principal + Interest = P + I = 16,000 + 13,000 = \$29,000.

The man will pay back \$29,000 to the lender.

Example 2: If \$450 amounts to \$504 in 3 years, what will \$615 amount to in $2\frac{1}{2}$ years?

=> Given that $A = 504$, and $P = 450$, we have $SI = A - P = 504 - 450 = 54$.

=> $SI = \dfrac{P \times r \times t}{100}$ => $54 = \dfrac{450 \times r \times 3}{100}$ => $r = \dfrac{54 \times 100}{450 \times 3} = 4\%$

=> Again, SI for \$615 for $2\frac{1}{2}$ years = $\dfrac{615 \times 4 \times \frac{5}{2}}{100}$ = \$61.50

=> Amount = $A = P + I = 615 + 61.50 = \$676.50$.

So, \$615 will amount to \$676.50 in $2\frac{1}{2}$ years.

3.9.2 Compound Interest

In the case of simple interest, the interest for the first period of time, the second period of time, the third period of time, and onwards remains the same. However, if the borrower and the lender agree that the amount at the end of the first period of time will become the principal for the second period of time, and the amount at the end of the second period of time will become the principal for the third period of time, and so on, then this arrangement is called **compounding**. It is called compounding because for every successive period of time, the principal is greater than that of the previous period of time, or, in other words, the interest gets compounded.

Let's see an example if the conditions of lending are (a) simple interest and (b) compound interest.

Example 3: A man borrowed \$10,000 at a rate of 10% per annum for 3 years. How much shall he pay back to the lender if the interest payable is (a) simple interest and (b) compound interest?

Year	Simple interest	Compound interest
First	$SI = 10,000 \times 10\% =$ \$1,000	$CI = 10,000 \times 10\% = \$1,000$; $A = P + I = 10,000 + 1000 = \$11,000$; $I = CI$ for the first period of time.
Second	$SI = 10,000 \times 10\% =$ \$1,000	$CI = 11,000 \times 10\% = \$1,100$; Amount of the previous year = Principal of the current year; $A = P + I = 11,000 + 1,100 = \$12,100$; $CI > SI$ for the second period of time.
Third	$SI = 10,000 \times 10\% =$ \$1,000	$CI = 12,100 \times 10\% = \$1,210$; Amount of the previous year = Principal of the current year; $A = P + I = 12,100 + 1,210 = \$13,310$; CI for the third period of time > CI for the second period of time.

	Total $SI = \$3,000$	Total $CI = 13,310 - 10,000 = \$3,310$
	Amount = $13,000	**Amount = $13,310**

*If the question does not specify whether the interest is compound interest or simple interest, you should consider it as simple interest.

Formula:

$A = P\left(1 + \frac{r}{100}\right)^n$; where A = Amount, P = Principal, r = Rate of interest, and n = Period

The above question can be solved with the help of this formula.

$$A = P\left(1 + \frac{r}{100}\right)^n$$

$= 10,000\left(1 + \frac{10}{100}\right)^3 = 10,000\left(1 + \frac{1}{10}\right)^3 = 10,000\left(\frac{11}{10}\right)^3 = \frac{10,000 \times 11 \times 11 \times 11}{10 \times 10 \times 10} =$ $\$13,310.$

So, $CI = A - P = 13,310 - 10,000 = \$3,310.$

3.9.2.1 Half-Yearly and Quarterly compounding

We have been referring to time as a period. The time is not necessarily measured in years only. The compounding of interest can be half-yearly (6 months), quarterly (3 months), or even monthly.

Half-yearly compounding means the interest would be calculated two times in a year; similarly, quarterly compounding means the interest would be calculated four times in a year, and like-wise, monthly.

Example 4: A man borrowed $10,000 at a rate of 10% per annum for one year. How much shall he pay back to the lender if the interest payable is calculated on a (1) half-yearly compounding and (2) quarterly compounding basis?

(1) **Half-yearly compounding:**

=> $P = 10,000, r = 10\%$ p.a., $t = 1$ year

Step 1: Covert rate of interest $r = 10\%$ p.a. into half yearly rate, so effective $r = 10\%/2 = 5\%$ per 6-months

Step 2: Covert time in years into periods; $n = t$ years $= 1$ year $\times\ 2$ = two periods of 6-months

Step 3: Apply CI formula

$A = P\left(1 + \frac{r}{100}\right)^n = 10,000\left(1 + \frac{10/2}{100}\right)^{1\times 2} = 10,000\left(1 + \frac{1}{20}\right)^2 = 10,000\left(\frac{21}{20}\right)^2 =$ $\$11,025$.

(2) **Quarterly compounding:**

=> $P = 10,000$, $r = 10\%$ p.a., $t = 1$ years, $n = 1 \times 4 = 4$

$A = P\left(1 + \frac{r}{100}\right)^n = 10,000\left(1 + \frac{10/4}{100}\right)^{1\times 4} = 10,000\left(1 + \frac{1}{40}\right)^4 = 10,000\left(\frac{41}{40}\right)^4 =$ $\$11,038$.

As you see that the amount after compounding the sum on a quarterly basis ($11,038) is more than that on a half-yearly basis ($11,025), we can conclude that as the period of compounding increases, the interest increases.

3.9.2.2 Calculating Rate of Interest and Time

Example 5: If a sum compounding annually doubles itself in 4 years, what is the rate of interest?

(A) 9.89%

(B) 18.92%

(C) 24.89%

Say $P = 100$, so $A = 200$, and $n = 4$ years.

$=> 200 = 100\left(1 + \frac{r}{100}\right)^4 => 2 = \left(1 + \frac{r}{100}\right)^4;$

$=> 2^{\frac{1}{4}} = 1 + \frac{r}{100}$ taking fourth root of both sides;

$=> 1.1892 = 1 + \frac{r}{100}$

$=> 1.1892 - 1 = \frac{r}{100}$

$=> 0.1892 = \frac{r}{100}$

$=> r = 0.1892 \times 100 = 18.92\%$

The correct answer is option B.

Alternate approach:

Remember that if a sum of money P doubles at interest r and time n under compound interest, then an approximate result may be used: $r \times n = \approx 72$ (provided r is small).

Here, $r \times n =\approx 72$, where $\boldsymbol{n} = 4$ (given). Thus, $\boldsymbol{r} =\approx 72/4 =\approx 18\%$.

Example 6: If a sum compounding annually becomes 1.331 times of itself at a rate of 10%, what is the time period in years?

(A) 2%

(B) 3%

(C) 4%

Say $P = 100$, so $A = 1.331 \times 100 = 133.1$, and $r = 10\%$.

=> $133.1 = \left(1 + \frac{10}{100}\right)^n$ => $1.331 = (1.1)^n$; again by hit and trial, we can deduce that $n = 3$ years.

Note: Difference between CI and SI for 2 years equals the SI on the "SI amount for a year",
i.e. $r\%$ of $\left(\frac{Pr}{100}\right) = \left(\frac{r}{100}\right) \times \left(\frac{Pr}{100}\right)$

3.9.2.3 Population Compounding

The concept of compounding can be applied to calculate the population compounding at a certain rate.

Example 5: If the population of a town increases at 6% p.a., but decreases due to emigration by 1% p.a., what is the net percentage increase in the population in 3 years?

=> Say the population before the increase is P_0 and after 3 years is P_3

So, $P_n = P_0\left(1 \pm \frac{r}{100}\right)^n$

Net increase in population = r = Growth rate - Emigration rate = 6 - 1 = 5%

$P_3 = 100\left(1 + \frac{(6-1)}{100}\right)^3 = 100\left(1 + \frac{1}{20}\right)^3 = 100\left(\frac{21}{20}\right)^3 = 100 \times 2.1 \times 2.1 \times 2.1 = 115.76;$

So, the net % increase $= 115.76 - 100 = 15.76\%$.

Different rates of compounding for different periods:

So far, we have seen that there is one rate of compounding for the whole period. However, there can be two or three different rates for different periods. Let us see how.

Example 6: If the population of a town increases at 5% for the first 2 years, increases at 2% for the next 3 years, and then decreases at 2% for the next 2 years, what is the net percentage increase in population in 7 years?

The formula of compounding can be modified as:

$$P_n = P_0\left(1 \pm \frac{r_1}{100}\right)^{n_1}\left(1 \pm \frac{r_2}{100}\right)^{n_2}\left(1 \pm \frac{r_3}{100}\right)^{n_3};$$

where $n_1, n_2,$ & n_3 are the time periods, $n = n_1 + n_2 + n_3$; and $r_1, r_2,$ & r_3 are the rates for respective time periods.

$$P_7 = 100\left(1 + \frac{5}{100}\right)^2\left(1 + \frac{2}{100}\right)^3\left(1 - \frac{2}{100}\right)^2 = 100\left(\frac{21}{20}\right)^2\left(\frac{51}{50}\right)^3\left(\frac{49}{50}\right)^2 = 112.37$$

So, the net % increase = 112.37 - 100 = 12.37%.

3.10 Expressions

Any form of numeric representation representing a number is called an **Expression.**

Example: $-3, (x - 2y), \frac{2p - 3q^3}{9}, \sqrt[3]{\frac{y^6}{2 - y^2}}$. All these expressions including, -3, show numbers. When we plug in or key in the value of variables x, y, p, q in the expressions, we get their numeric values.

An expression does not contain '=' sign. If two expressions are equated by '=' sign, it is called an **Equation**; however if they are separated by either of $<, >, \geq, \leq$ signs, it is called an **Inequation** or **Inequality.**

Simplification of expressions:

Combining like terms: $(2x + 3x)$ is an expression in a distributive or an expanded form. We can simplify the expression by adding like terms. So, the simplified form is: $(2x + 3x) = 5x$.

Expanded form of expression	Simplified form of expression	Comments
$2x^2 + 3x - 4x - 4$	$2x^2 - x - 4$	$3x$ and $-4x$ are like terms, and $2x^2$ and 4 unlike terms.
$(3x - 4) - (4y - 4)$	$3x - 4 - 4y + 4$ $= (3x - 4y)$	$3x$ and $-4y$ are not like terms. (Apple would be added to apple & not to orange.)
$2x^2 + 3\sqrt{x^4} - 4x$	$2x^2 + 3\sqrt{x^4} - 4x$ $= 2x^2 + 3x^2 - 4x$ $= 5x^2 - 4x$	Simplify the exponents.
$\frac{2x}{y} - \frac{x}{y}$	$= \frac{x}{y}$	Consider $\frac{x}{y}$ as one variable.
$4x^2y - 10xy^2$	$4x^2y - 10xy^2$ $= 2xy(2x - 5y)$	Take out common factor. Depending on the situation you may have to either expand an expression or simplify it.

3.11 Equations

Equality of two expressions is called Equation.

Example: $2x + 3 = 0$ is an equation.

Here there are two expressions: '$2x + 3$' and '0'. An equation has two sides: Left had side $(2x + 3)$ and right hand side (0). The left had side (LHS) always equals the right hand side (RHS).

Manipulating an equation:

A. Whatever you do to any side of an equation, do the same to the other side:

Say the equation is $2x + 3 = 0$ and we are interested in deducing the value of the variable x. To get the value of x, we must try to isolate x on either side of the equation.

Here x accompanies a co-efficient '2', and a number, 3, added to it. So, we must get rid of '3' and '2'. Let us see how.

$2x + 3 = 0$

=> $2x + 3 - 3 = 0 - 3$; Subtracting '3' from both the sides: LHS and RHS

=> $2x = -3$;

=> $\frac{2x}{2} = -\frac{3}{2}$; Dividing both the sides by '2' to get rid of x's co-efficient '2'

$\boldsymbol{x = -\frac{3}{2}}$; We finally get the value of x.

Example 1: Find the value of x for the equation: $\frac{3x-7}{5} = 7$.

$\frac{3x-7}{5} = 7$

=> $\frac{3x-7}{5} \times 5 = 7 \times 5$; Multiplying both the sides by '5' to get rid of '5' in the denominator

=> $3x - 7 = 35$;

=> $3x - 7 + 7 = 35 + 7$; Adding '7' on both the sides to get rid of '-7' from LHS, we get

$3x = 42$;

=> $\frac{3x}{3} = \frac{42}{3}$; Dividing both the sides by '3' to get rid of x's co-efficient '3'

$\boldsymbol{x = 14}$

B. Transposition:

The above method of manipulation is time-consuming. Let us see another method to get the same result: **Transposition.**

We take the same equation: $2x + 3 = 0$. We can transpose '3' from LHS to RHS.

"*If any expression on any side is positive, it will be transposed as negative on the other side and vice-versa.*"

$2x + 3 = 0$

=> $2x = -3$; '+3' is transposed from LHS to RHS as '-3'; Note the sign change!

$\boldsymbol{x = -\frac{3}{2}}$; Co-efficient '2' transposed from LHS to RHS as '$\frac{1}{2}$'; Note no sign change

Example 2: Find the value of x for the equation: $\frac{3x-7}{5} = 7$.

=> $\frac{3x-7}{5} = 7$

$3x - 7 = 7 \times 5$; Cross-multiplication of '5'. Since '5' was on the denominator of the LHS, it will go to RHS as multiplicand. Remember no sign change.

$3x - 7 = 35$;

$3x = 35 + 7$; '-7' is transposed from LHS to RHS as '+7'; Remember sign change

$3x = 42$;

$x = \frac{42}{3}$; Co-efficient '3' transposed from LHS to RHS as '$\frac{1}{3}$'; Remember no sign change

$\boldsymbol{x = 14}$

Example 3: Find the value of x for the equation: $7\left(\frac{x}{2} + 3\right) = 21$

=> $7\left(\frac{x}{2} + 3\right) = 21$

$\frac{x}{2} + 3 = \frac{21}{7}$; Multiplicand '7' transposed from LHS to RHS as '$\frac{1}{7}$'; Remember no sign change

$\frac{x}{2} + 3 = 3$

$\frac{x}{2} = 3 - 3$; '+3' is transposed from LHS to RHS as '-3'; Remember sign change

$\frac{x}{2} = 0$;

$x = 0 \times 2$; Cross-multiplication of '2'. Since '2' was on the denominator of the LHS, it will go to RHS as multiplicand. Remember no sign change.
$\boldsymbol{x = 0}$

Example 4: Find the value of x for the equation: $3\left(\sqrt{x^2+3}\right) = 9$.

$=> 3\left(\sqrt{x^2+3}\right) = 9$

$\sqrt{x^2+3} = \frac{9}{3} => \sqrt{x^2+3} = 3$; Transposing '3'.

$=> (\sqrt{x^2+3})^2 = 3^2 => x^2 + 3 = 9$; Squaring both the sides to get rid of square-root

$=> x^2 = 9 - 3 => x^2 = 6$; Transposing '3'.

$x^2 = 6$;

$=> \sqrt{x^2} = \pm\sqrt{6}$; taking square-root of both the sides to get rid of square.

$\boldsymbol{x = \pm\sqrt{6}}$

Example 5: Find the value of x for the equation: $\dfrac{3(x-2)+2(3x-2)}{x-5} = -2$.

$=> \dfrac{3(x-2)+2(3x-2)}{x-5} = -2$

$=> 3(x-2) + 2(3x-2) = -2(x-5)$; Cross-multiplication

$=> 3x - 6 + 6x - 4 = -2x + 10$; Expansion

$=> 3x + 6x + 2x = 6 + 4 + 10$; Transposing like terms and isolating x

$=> 11x = 20$; Adding like terms

$\boldsymbol{x = \dfrac{20}{11}}$

3.11.1 System of Equations

So far we have dealt with one variable and one equation. If we have two or more equations, we call them **Systems of Equations.** You can get the unique value of the variable, say x, from one equation; however if you have to deal with more than two variables, you have to deal with more than two equations.

Say the equation is $2x + 3y = 13$. Since there are two variables, x and y, we need two equations to get their unique values. Say another equation is $2x - 3y = -5$.

So, we have a system of equations: $2x + 3y = 13$ and $2x - 3y = -5$. There are quite a few methods of solving the system of equations; we discuss two of them.

'System of equations' is also referred to as **'Simultaneous Equations'**.

Linear equation:

When the exponent or the power of variable(s) in an equation is '1', it is called a linear equation. $2x + 3y = 13$ and $2x - 3y = -5$ are linear equations.

The standard form of simultaneous equations is $ax + by = c$; where x and y are the coefficients of the variables x and y respectively and c is a constant term.

Solving simultaneous equations:

1. Method of Elimination:

In this method, we eliminate one of the variable using both the equations and get the numeric value of other variable.

By plugging in the numeric value of the other variable in any equation, we get numeric value of the first variable. Follow the following steps to solve the equations.

Step 1: Simplify and transform both the equations in the form $ax + by = c$, if needed

=> The two equations ($2x + 3y = 13$ and $2x - 3y = -5$), we are dealing with are already in the standard form.

Step 2: Decide which variable should be eliminated

=> If you are interested in finding the value of x, then eliminate y, and vice-versa; however, if you need to calculate the values of both the variables, decide which variable is relatively easier to eliminate.

=> If the co-efficients of one of the two variables are same in the equations, you should eliminate that variable; however, if the co-efficients are not same, you may still decide which variable should be eliminated; it depends on how easy it is for you to make the absolute value of the co-efficients of a variable equal. This part can be better understood in step 3.

For our set of equations: $2x + 3y = 13$ and $2x - 3y = -5$, we have to find out the values of both x and y. We decide that we will eliminate variable y as the coefficients of y for the equations are equal and are of opposite sign. This part can be better understood in step 4.

Step 3: Make the co-efficients of the variable to be eliminated equal

=> If the co-efficients of the variable to be eliminated are not same in both the equations, you should make them equal by multiplying with minimum possible numbers to the co-efficient(s).

> For our set of equations: $2x + 3y = 13$ and $2x - 3y = -5$, we have already decided that we will eliminate variable y, since the absolute value of the co-efficients of y in both the equations are equal ('3'), we need not multiply equation(s) with a number(s) to make co-efficients equal.

Step 3: Add or subtract the equations

=> If the signs of the co-efficients of the variable to be eliminated are same, add the equations, else subtract.

> For our set of equations: $2x + 3y = 13$ and $2x - 3y = -5$, since the signs of the co-efficients of y in both the equations are same (–), we should add the equations.
>
> $2x + 3y = 13$ ————————(1); name it eqn. (1)
> $2x - 3y = -5$ ————————(2); name it eqn. (2)
> ——————————
>
> **$4x = 8$**; $+3y$ and $-3y$ get cancelled.
>
> $x = 8/4$ => **$x = 2$**.

Step 4: Calculate the value of other variable

=> Substitute the value of the variable derived in step 3 in any of the equations convenient to you.

> Let us plug in the value of **$x = 2$** in the eqn (1), and we get:
>
> $2.2 + 3y = 13 \Rightarrow 3y = 13 - 4 = 9 \Rightarrow 3y = 9 \Rightarrow y = \frac{9}{3}$
>
> => **$y = 3$**.
>
> So, **$x = 2$**, and **$y = 3$**.

Step 5 (Optional): Cross check the values derived by plugging in both the equation

=> Plug in the values of both the variables in both the equations; LHS and RHS must be equal.

Eqn(1): $2x + 3y = 13 \Rightarrow 2(2) + 3(3) = 13 \Rightarrow 4 + 9 = 13 \Rightarrow \mathbf{13 = 13}$.

Eqn(2): $2x - 3y = -5 \Rightarrow 2(2) - 3(3) = -5 \Rightarrow 4 - 9 = -5 \Rightarrow \mathbf{-5 = -5}$.

This proves that the values derived are correct.

Example 1: Find out the values of x and y for the equations $3x + 5y - 21 = 0$ and $2x + 3y = 13$.

First, make the equation $3x + 5y - 21 = 0$ in the standard form: $3x + 5y = 21$. Second equation $2x + 3y = 13$ is already in standard form.

=> So the equations are:

$3x + 5y = 21$————-(1)
$2x + 3y = 13$————-(2); name the equations

Since we have to find out the values of both the variables x and y, it is immaterial to choose which variable should be eliminated; however looking at the equations, we decide that we will eliminate x.

Next task is to make the co-efficients of x in both the equations equal.

By multiplying '2' to the eqn(1), and '3' to the eqn(2), we get,

$2 \times (3x + 5y = 21)$ ————-(1)
$3 \times (2x + 3y = 13)$ ————-(2);

=> $6x + 10y = 42$————-(1)
=> $6x + 9y = 39$————-(2);

=> Since the signs of the co-efficients of x for both the equations are same, we will subtract eqn (2) from eqn (1).

$6x + 10y = 42$————-(1)

$6x + 9y = 39$————-(2);
–   –   –

————————————

When we subtract eqn (2) from eqn (1), the signs of the co-efficients and the constant of eqn (2) will change (+ will change to – and vice-versa)

=> After cancellation of $6x$, we get $10y - 9y = 42 - 39$ => $\boldsymbol{y = 3}$.

=> Now, plug in the value of y in any equation; we choose eqn (2)

$2x + 3y = 13 \Rightarrow 2x + 3.3 = 13 \Rightarrow 2x + 9 = 13 \Rightarrow 2x = 13 - 9 \Rightarrow 2x = 4$

=> $\boldsymbol{x = 2}$

So, we get $\boldsymbol{x = 2}$, and $\boldsymbol{y = 3}$.

2. Method of Substitution:

In this method, by choosing an equation we calculate the value of a variable in terms of other variable and constant, plug in that variable's value in the second equation, and get the numeric value of the other variable.

Again, by plugging in the numeric value of the other variable in any equation, we get the numeric value of the first variable. Follow the following steps to solve the equations.

Step 1: Decide which variable should be substituted

=> We will deal with the two equations ($x + 3y = 11$ and $2x - 3y = -5$).
=> Observe which variable in an equation does not have its co-efficient; that very variable should be chosen to be substituted; the reason for choosing that variable is that though its value would be in terms of the other variable and the constant, it would not have any denominator to deal with.

On this basis, we choose eqn (1): $x + 3y = 11$, and variable x to be substituted in eqn (2): $2x - 3y = -5$

Step 2: Calculate the value of the chosen variable in terms of the other variable and constant

$x + 3y = 11 \Rightarrow \boldsymbol{x = 11 - 3y}$——— say equation (1)

Step 3: Substitute the value calculated in step 2 in the other equation

By plugging in the value of $\boldsymbol{x = 11 - 3y}$ in the eqn (2): $2x - 3y = -5$, we get,

$2.(11 - 3y) - 3y = -5 \Rightarrow 22 - 6y - 3y = -5 \Rightarrow -9y = -5 - 22 \Rightarrow -9y = -27$

$$=> y = \frac{-27}{-9}$$

$=>$ $\boldsymbol{y = 3.}$

Step 4: Calculate the value of other variable

Substitute the value of the variable derived in step 3 in any of the equation convenient to you, preferably equation (1).

Let us plug in the value of $y = 3$ in the eqn: $x = 11 - 3y$, and we get:

$x = 11 - 3 \times 3 => x = 11 - 9 => x = 2.$

So, $\boldsymbol{x = 2}$, and $\boldsymbol{y = 3}$.

Example 2: Find out the values of x, & y from the equations: $3x + 5y - 21 = 0$ and $2x + 3y = 13$.

=> No variable in any equation is without co-efficient, so you may choose any variable and any equation to start with.

=> Say we chose eqn (2) and variable x

So, $2x + 3y = 13 => 2x = 13 - 3y => \boldsymbol{x} = \frac{13-3y}{2}$

=> By plugging in the value of $x = \frac{13-3y}{2}$ in eqn(1): $3x + 5y - 21 = 0$, we get,

$$3 \times \left(\frac{13 - 3y}{2}\right) + 5y - 21 = 0 => \frac{3(13 - 3y) + 2 \times 5y - 2 \times 21}{2} = 0$$

$$3(13-3y)+2\times5y-2\times21 = 2\times0 => 39-9y+10y-42 = 0 => y-3 = 0 => \boldsymbol{y = 3}.$$

=> By plugging in the value of $y = 3$ in eqn: $x = \frac{13-3y}{2}$, we get

$$x = \frac{13 - 3 \times (3)}{2} = \frac{13 - 9}{2} = \frac{4}{2} => \boldsymbol{x = 2}$$

So, $\boldsymbol{x = 2}$ and $\boldsymbol{y = 3}$.

Number of solutions

For a linear equation: $2x + 3y = 12$, we cannot find out the unique solution or the unique values of the variables x and y; however we can find out infinite number of consistent solutions. Say $x = 3$, then $y = 2$. Following illustration shows few consistent solutions.

Equation: $\mathbf{2x + 3y = 12}$

x	0	6	1	−1	3	2
y	4	0	10/3	14/3	2	8/3

For integer only solutions, we can follow the following method:

Find out one possible integer solution for the equation $2x + 3y = 12$: it is obvious that it is $x = 6$ and $y = 0$.

To get the next value of x, subtract the coefficient of y (with sign) from the value of x obtained and to get the next value of x, add the coefficient of x (with sign) to the value of y obtained previously. Thus gives us $x = 6 - 3 = 3$ and $y = 0 + 2 = 2$.

Alternatively, to get the next value of x, add the coefficient of y (with sign) to the value of y obtained and to get the next value of y, subtract the coefficient of x (with sign) from the value of y obtained previously. Thus gives us $x = 6+3 = 9$ and $y = 0-2 = -2$.

The above steps can be continued indefinitely to get all sets of values of integer values of x and y.

If only positive values are needed, we continue the process only till when the values of the variables remain positive.

Please remember that the above method is only applicable for two variables.

Unique vs. Consistent solution:

Let us see the following system of equations:

$2x + 3y = 12$———–(1) &
$4x + 6y = 24$———–(2)

You would have observed that eqn (2) is derived from the parent eqn (1) by multiplying itself by '2', so in fact both the equations are the same. So, though these equations are **consistent**, they would **not have unique solution** rather **infinite number of solutions.**

Let us see the following system of equation:

$2x + 3y = 12$———–(1) &
$4x + 6y = 4$ ———–(2)

You would have observed that eqn (2) is derived from the parent eqn (1) by multiplying itself by '2' to its LHS, but its RHS is something different, other than anticipated '24', so the given equations are **inconsistent.** They **would not have any solution** at all.

Let us see the following system of equations:

$4x + 6y = 24$———-(1) &
$3x + 2y = 13$———-(2)

You would have observed that eqn (2) is different from the parent eqn (1), and you cannot derive eqn (2) by manipulating eqn (1), so these equations given are **consistent**, and **would have unique solution**.

Following table illustrates the nature of equations and number of solutions.

$\boldsymbol{a_1x + b_1y = c_1}$———(1) & $\boldsymbol{a_2x + b_2y = c_2}$———(2)

Ratio of co-efficients and constant	Nature of equation	Numbers of solution(s)
$\frac{a_1}{a_2} = \frac{b_1}{b_2} = \frac{c_1}{c_2}$	Consistent equations	Infinitely many solutions
$\frac{a_1}{a_2} = \frac{b_1}{b_2} \neq \frac{c_1}{c_2}$	Inconsistent equations	No solution
$\frac{a_1}{a_2} \neq \frac{b_1}{b_2}$	Consistent equations	Unique solution

There may be two identical equations which may seem to look unique equations, for example:

$2x + 3y = 12$———(1) &

$\frac{x + \frac{3}{2}y}{6} = 1$———-(2).

If you did not pay attention to this, you may fall in the trap laid by the test-maker, and without solving the equations, you may decide that the two equations will render unique values of x and y.

3.11.2 Quadratic Equation

A linear equation has variable(s) which have the exponent or power of '1'; whereas a **Quadratic Equation** comprises of at least one variable which has the exponent or power of '2'. The name Quadratic comes from "Quad" meaning "Square", as the variable(s) get squared.

Standard form of a quadratic equation is: $ax^2 + bx + c = 0$, where $a \neq 0$. There are at the most two values of x, satisfying the equation; these values are also called **Roots** of the equation.

Methods of calculating roots:

1. Factoring quadratic equation:

Say a quadratic equation is: $2x^2 + 10x + 12 = 0$. Values: $x = -2$ and $x = -3$ satisfy the equation, so $x = -2$ and $x = -3$ are the roots of the equation.

We will find out the roots of the equation through factoring method.

Step 1: Manipulate the equation to make it in standard form: $ax^2 + bx + c = 0$

=> Our eqn is $2x^2 + 10x + 12 = 0$. We simplify the equation by dividing it by '2', so we get $x^2 + 5x + 6 = 0$. Here $a = 1, b = 5$, and $c = 6$.

Step 2: Split "ac"

=> Calculate the product of a and c.

=> For our eqn: $\boldsymbol{ac = 1 \times 6 \Rightarrow ac = 6}$.

=> Mentally run through the factors of the $\boldsymbol{ac}$, and choose any two of them, such that their sum equals **'co-efficient of $\boldsymbol{x : b}$.'**

=> In our equation: $x^2 + 5x + 6 = 0$, 'co-efficient of $x : b = 5$', and $ac = 6$. Factors of '6' are {1, 2, 3, 6}.

There are two possible pairs of factors that make their product equal to $6 : 1 \times 6 = 6$, and $2 \times 3 = 6$; but out of these two pairs, we must choose {2, 3} as their sum also makes co-efficient of $x : b = 2 + 3 = 5$).

So the equation can be rewritten as: $x^2 + 2x + 3x + 6 = 0$.

We can pair first two terms and last two terms as: $x(x + 2) + 3(x + 2) = 0$. Here we can pull out $(x + 2)$ as common, so the equation becomes the product of two factors:

$(x + 2)(x + 3) = 0.$

We know that if the product of two expression is '0', at least one of the expressions must be '0'; so, either $x + 2 = 0$ or $x + 3 = 0$. This gives either $x = -2$ or -3. There are the two roots or the solution of the equation.

2. Using formula:

Roots of the equation: $ax^2 + bx + c = 0$ are given by:

$$x_1 = \frac{-b + \sqrt{b^2 - 4ac}}{2a}; \&$$

$$x_2 = \frac{-b - \sqrt{b^2 - 4ac}}{2a}$$; where $\boldsymbol{b^2 - 4ac}$ is called discriminant (**Δ**)

The equation $2x^2 - 2x - 12 = 0$ can be simplified as $x^2 - x - 6 = 0$; where $a = 1, b = -1$, and $x = -6$. The roots are referred to as $x_1 \& x_2$.

$$x_1 \ \& \ x_2 = \frac{-b \pm \sqrt{b^2 - 4ac}}{2a} = \frac{-(-1) \pm \sqrt{(-1)^2 - (4 \times 1 \times -6)}}{2 \times 1} = \frac{1 \pm \sqrt{1 + 24}}{2} = \frac{1 \pm 5}{2}$$

$$x_1 = \frac{1 + 5}{2} = 3, \text{ or } x_2 = \frac{1 - 5}{2} = -2; \text{ so } x_1 = -2 \text{ and } x_2 = 3$$

The term discriminant (**Δ**) = $\boldsymbol{b^2 - 4ac}$ can be positive/Negative/Zero, depending on the values of a, b, and c.

The following table illustrates the relationship between $\boldsymbol{\Delta = b^2 - 4ac}$, and two roots of the equation.

$\Delta = b^2 - 4ac$	Nature of roots	Form/Value or Example
Positive (> 0)	Real & Unequal; Two distinct roots	x_1 & $x_2 = \dfrac{-b \pm \sqrt{b^2 - 4ac}}{2a}$
	If Δ is perfect square, roots are Rational	For $x^2 + 5x + 6 = 0$; $x = -2$, or $= -3$
	If Δ is not a perfect square, roots are Irrational	For $x^2 + 2x - 6 = 0$; $x_1 = (-1 + \sqrt{7})$, $x_2 = (-1 - \sqrt{7})$
Equal (=)	Real; only one root	$-\dfrac{b}{2a}$; For $x^2 + 4x + 4 = 0; x = -2$
Negative (< 0)	Imaginary roots	$a + ib$; For $x^2 + 2x + 6 = 0$, $x_1 = (-1 + \sqrt{-5})$, $x_2 = (-1 - \sqrt{-5})$

3.11.2.1 Roots of quadratic equation

So far we have seen how we can find out the roots of the quadratic equation. There may be instances in which you have to form a quadratic equation if the values of roots are given. Let us see how we can do it.

We have seen that the factors of $x^2 + 5x + 6 = 0$ are $(x + 2)$ and $(x + 3)$; where the roots are: $x = -2$ or $x = -3$. So, we can write the equation as $[x - (-2)][x - (-3)] = 0$ or $(x - x_1)(x - x_2) = 0$.

If you expand $(x - x_1)(x - x_2) = 0$, you will get $x^2 - (x_1 + x_2) + x_1x_2 = 0$; where $x_1 + x_2$ = sum of roots, and x_1x_2 = product of roots.

$$\mathbf{x^2 - (Sum\ of\ Roots)x + (Product\ of\ Roots) = 0}$$

Let us call it **Normal form** of quadratic equation. Here **co-efficient of *x*** is the **middle term: 'Negative of Sum of roots'** and the **constant** is the **last term: 'Product of roots'.**

Standard form of equation: $ax^2 + by^2 + c = 0$ can be written as $x^2 + \left(\frac{b}{a}\right) + \left(\frac{c}{a}\right) = 0$

=> $x^2 - \left(-\frac{b}{a}\right) + \left(\frac{c}{a}\right) = 0$. So, **Sum of roots =** $-\frac{b}{a}$ and **Product of roots =** $\frac{c}{a}$.

Example 1: Two roots of a quadratic equation are $-\frac{1}{2}$ and $\frac{3}{2}$. Find out the equation.

Sum of the roots = $-\frac{1}{2} + \frac{3}{2} = 1$, and product of roots = $-\frac{1}{2} \times \frac{3}{2} = -\frac{3}{4}$

=> So, the equation is: $x^2 - (1)x + \left(-\frac{3}{4}\right) = 0$;

=> $x^2 - x - \frac{3}{4} = 0$

=> $\mathbf{4x^2 - 4x - 3 = 0}$

Example 2: One of the roots of a quadratic equation is 2 and the product of roots is 6. Find out the equation.

Product of roots = $x_1x_2 = 2 \times x_2 = 6 => x_2 = 3$;

=> Now, Sum of the roots = 2 + 3 = 5;

=> So, the equation is: x^2 – (Sum of roots)x + (Product of roots) = 0.

=> $\mathbf{x^2 - 5x + 6 = 0}$

Example 3: For a quadratic equation: $2x^2 - 4x = 3$, find out the sum of roots, and product of roots.

The equation can be written in standard form as: $2x^2 - 4x - 3 = 0$

=> The normal form requires that the co-efficient of x^2 must be '1' so, we divide the equation by '2', and we $x^2 - 2x - \frac{3}{2} = 0$

=> The co-efficient of $x = -2$; so the sum of roots = 2, and the constant = $-\frac{3}{2}$; so the product of roots = $-\frac{3}{2}$.

3.11.2.2 Higher order equations

Though higher order equations/polynomial equations are out of scope, there may an equation such as $\mathbf{x^3 - 5x^2 + 6x = 0}$, which is a not a quadratic equation; however you can still solve the equation in a quadratic equation way. Let us see how.

$\mathbf{x^3 - 5x^2 + 6x = 0}$**;**

=> $x(x^2 - 5x + 6) = 0$

=> $x(x - 2)(x - 3) = 0$

=> Either $x = 0$ or $x - 2 = 0$ or $x - 3 = 0$ => $\mathbf{x = 0}$ or **2** or **3**. So there are three roots of the equation.

Never divide the equation by x if you are not sure that $x \neq 0$. Not understanding this concept may land you in trouble. See the following equation.

$\mathbf{x^3 - 4x^2 + 4x = 0}$;

=> $x(x^2 - 4x + 4) = 0$ => Dividing the equation by x, we get $x^2 - 4x + 4 = 0$

=> $(x - 2)^2 = 0$ => $x = 2$, a unique value of x! Right! No it's wrong.

We have missed out $x = 0$; so there are two possible values of x: 0 and 2.

3.12 Absolute numbers

We know that any real number can either be 0, positive or negative. When we ignore the negative sign of a number, we seem to be interested in its absolute value and not in its directed value.

In other words, absolute value of a number is its distance from '0'. If a number is presented between two-vertical bar (pipe ||) symbol, it means we are interested in its absolute value.

Say, a number is x, then its absolute value is $|x|$, as depicted in the figure below.

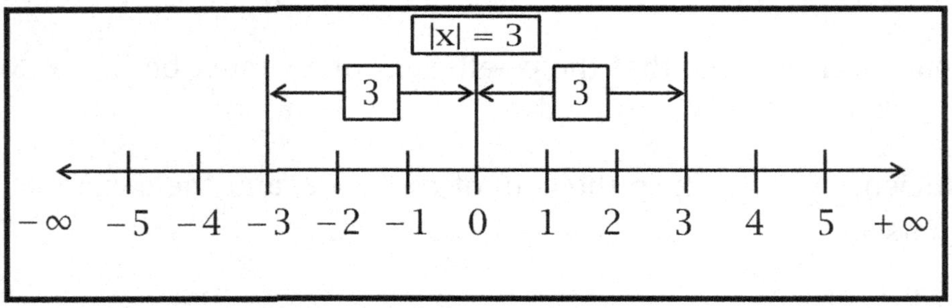

Key properties:

(1) $\sqrt{x^2} = |x|$

(2) $|\pm x| = |x|$

(3) $|x| = |-x|; -|x| \neq |-x|$

(4) $|x| \geq 0$

(5) $|x \pm y| = |y \pm x|$

Say $x = 10$ and $y = -10$, then
$|x - y| = |10 - (-10)| = |10 + 10| = 20$;
$|y - x| = |-10 - 10| = |-20| = 20$.

(6) $|x + y| \leq |x| + |y|$

Say $x = 10$ and $y = 10$, then
$|x + y| = |10 + 10| = 20$;
$|x| + |y| = |10| + |10| = 10 + 10 = 20$, hence $|x + y| = |x| + |y|$

But, say $x = 10$ and $y = -10$, then
$|x + y| = |10 + (-10)| = |10 - 10| = 0$;
$|x| + |y| = |10| + |-10| = 10 + 10 = 20$, hence $|x + y| < |x| + |y|$

This deduction is an important finding and is tested by test-makers.

Remember that the equality holds true only when both x and y are of the same sign and the inequality holds true when x and y are of opposite signs.

(7) $|x - y| \geq |x| - |y|$

Here too, the equality holds true only when both x and y are of the same sign and the inequality holds true when x and y are of opposite signs.

(8) $|x \times y| = |y \times x| = |x| \times |y| = |y| \times |x|$

(9) $|x \div y| = |x| \div |y|$

Example: If $7|2x - 3| = 21$, what is the value of x?

Simplify $7|2x - 3| = 21$ by dividing both the sides by '7'; we get $|2x - 3| = 3$.

=> $2x - 3 = 3$ or $-(2x - 3) = 3$.

=> $2x - 3 = 3$ => $2x = 6$ => $x = 3$; or

=> $-(2x - 3) = 3$ => $-2x + 3 = 3$ => $-2x = 0$ => $x = 0$.

So $x = 0$ or 3.

3.12.1 Inequality

To understand Inequality, first understand Equality.

$x = 5$ means that the value of the variable x equals 5. The inequality could be anything $x \neq 4$, $x < 6$ or $x > 4$.

When the two sides of an expression are not equal, it is called Inequality. Inequalities use symbols $<, >, \leq$, or $\geq$ to describe the relationship between two expressions, for example: $10 < 12, 12 > 8, x \leq 4, x + 2y \geq 4$, etc.

Understanding inequality:

Inequality	Meaning	Example
$x < 5$	x is less than 5	x can be any value less than 5 on the number line; it can be 0, 1, -155, or any other value up to negative infinity. Do not assume that x is an integer. x can even be 4.999. Always consider that x is a real number until stated that it is an integer.
$y > 10$	y is greater than 10	y can be any value greater than 10 on the number scale. It can be 10.0001, 10.56, 21, or any other value up to infinity.
$p \leq 10$	p is less than or equal to 10	p can be any value less than or equal to 10. It can be 10, 4, -2.3, or any other value up to negative infinity.
$q \geq 10$	q is greater than or equal to 10	q can be any value greater than or equal to 10 on the number line. It can be 10, 14, 23.5, or any other value up to infinity.
$-2 \leq x < 4$; x is an integer	x is greater than or equal to -2 but less than 4	$x = \{-2, -1, 0, 1, 2, 3\}$
$-2 > x \geq -4$; x is an integer	x is less than -2 and greater than or equal to -4	$x = \{-3, -4\}$

Representation of inequalities on number line:

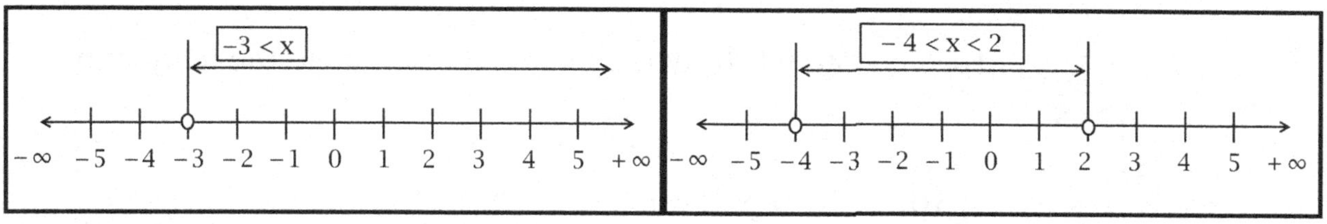

Key properties of Inequalities:

(1) You can **add or subtract anything from an inequality**.

Say, $x + 3 > 10$.

You can subtract 3 from both the sides, so the inequality becomes,
$x + 3 - 3 > 10 - 3 \Rightarrow x > 7$.

Take another example; say, $y - 3 > -10$.

You can add 3 on both the sides, so the inequality becomes,
$y - 3 + 3 > -10 + 3 \Rightarrow y > -7$.

(2) You can multiply or divide an inequality by a **positive** number.

Say, $2x > 10$.

You can divide the inequality by 2, so the inequality becomes,
$\frac{2x}{2} > \frac{10}{2} \Rightarrow x > 5$.

Say, $\frac{y}{3} > -10$. You can multiply the inequality by 3, so the inequality becomes,
$\frac{y}{3} \times 3 > -10 \times 3 \Rightarrow y > -30$.

(3) You can multiply or divide an inequality by a **negative** number, but doing so will ask for **reversing** the sign of inequality.

We know that $7 > 5$;

if we multiply an inequality by -1, it will become $-7 > -5$ which is wrong to say because $-7 < -5$; so when we multiply or divide an inequality with a negative number, we have to flip the sign.

Say, $-2x > 10$.

You can divide the inequality by -2, so the inequality becomes,

$\frac{-2x}{-2} \not> < \frac{10}{-2} => x < -5.$

Say, $\frac{-y}{3} > -10$. You can multiply the inequality by -3, so the inequality becomes,

$\frac{-y}{3} \times -3 \not> < -10 \times -3 => y < 30.$

3.12.2 Inequalities with absolute numbers

Few tough questions may be asked combining inequalities and absolute numbers.

Example 1: If $|x + 1| \leq 4$, what is x?

$|x + 1| \leq 4$

=> If $x + 1 \leq 4$ or $-(x + 1) \leq 4$

$x + 1 \leq 4 => x \leq 3$

Or, $-(x + 1) \leq 4 => x + 1 \geq -4 => x \geq -5$. (Pay attention to reversal of sign of inequality)

So, $\mathbf{3 \geq x \geq -5}$

It can be represented graphically as below.

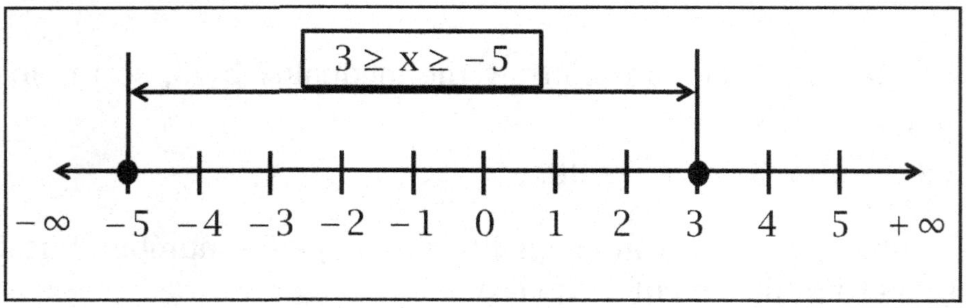

Example 2: If $|2x + 1| \geq 3$, what is x?

$|2x + 1| \geq 3 => 2x + 1 \geq 3$ or $-(2x + 1) \geq 3$

If $2x + 1 \geq 3 => 2x \geq 2 => x \geq 1$

Or, $-(2x + 1) \geq 3 => 2x + 1 \leq -3 => 2x \leq -4 => x \leq -2$. (Pay attention to reversal of sign of inequality)

So $\mathbf{-2 \geq x}$ or $\mathbf{x \geq 1}$

It can be represented graphically as below.

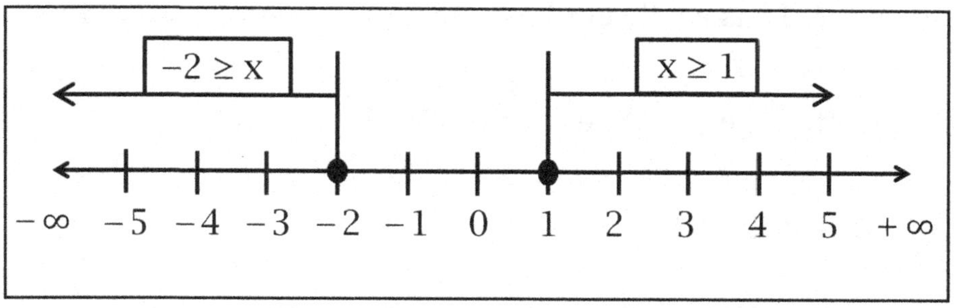

Example 3: If $8 + x^2 \geq 12$, what is the range of x?

$8 + x^2 \geq 12$

$\Rightarrow x^2 \geq 4$;

Taking square root of both the sides, we get,

$|x| \geq 2 \Rightarrow x \geq 2$ or $x \leq -2$.

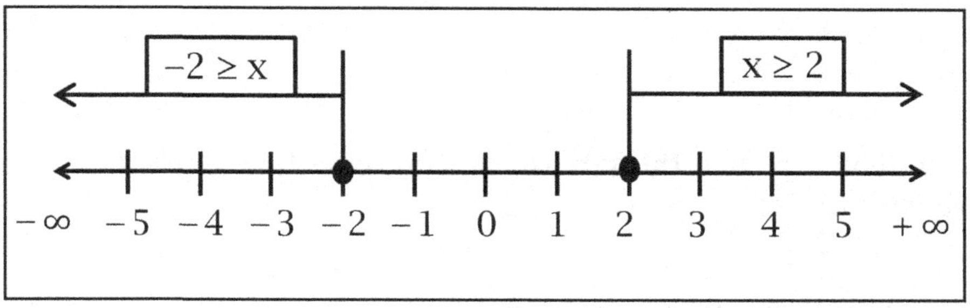

3.12.3 Compound Inequality

An inequality that has more than one inequality symbol in an expression is called compound inequality. $2 > x \geq 4$ is an example of compound inequality. GMAT inequality questions may involve more than one inequality. To solve such questions, you may have to transform several inequalities to a compound inequality.

Example 4: If $x > 4, x \leq 10$, and $x < 12$, what is the range of x?

Step 1: Fix 'more than x' or 'more than or equal to x' inequality.

There is only one 'more than x' inequality, so we fix $x > 4$.

Step 2: Fix 'less than x' or 'less than or equal to x' inequality.

There is one 'less than x' and one 'less than or equal to x' inequalities, but between the two inequalities $x \leq 10$, and $x < 12$, we choose the first one i.e. $x \leq 10$ since $x \leq 10$ is a subset of $x < 12$, not vice-versa; so we fix $x \leq 10$.

Step 3: Combine both the inequalities derived in steps 1 and 2.

Finally, we get $\mathbf{4 < x \leq 10}$.

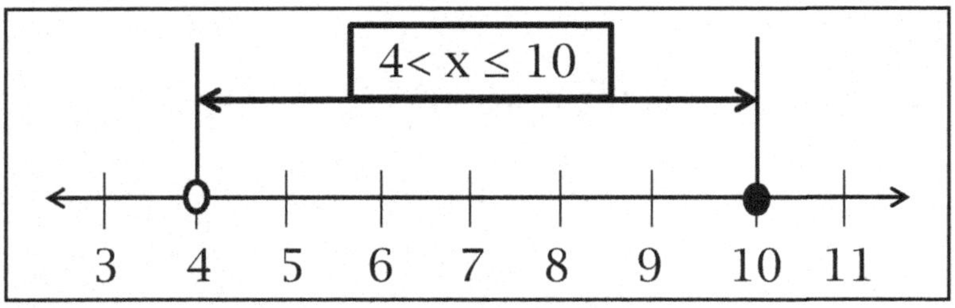

Note that:

If $x > y$, and $y > z$, then we can combine the inequalities as $x > y > z$ or $x > z$.

However,

If $x > y$, and $y < z$, then though we can combine the inequalities as $x > y < z$, but we **cannot** deduce that $x > z$.

3.12.4 Mathematical operations on inequalities

(1) **Addition and subtraction:**

If $x > y$ and $a < b$, we cannot deduce that $x + a < y + b$; it is wrong.

We must make both the inequalities in the same direction. So, we can rewrite the inequalities as $x > y$ and $b > a$, and now we can add these.

So, $\boldsymbol{x + b > y + a}$.

Let's take another example.

If $x - 3 < y \leq x + 6$, then we **cannot** write the inequality as $x < y \leq x + 6 + 3 \Rightarrow x < y \leq x + 9$. It is wrong!

If you subtract a number from one part of a compound inequality, you must subtract the same number from all the parts.

The simplified and corrected inequality is: $\boldsymbol{x < y + 3 \leq x + 9}$.

(2) **Multiplication and division:**

If $x < -\frac{y}{3} \leq 3x$, then we can simplify the inequality by multiplying it by -3.

So, $-3\left(x < -\frac{y}{3} \leq 3x\right) => -3x > (-\frac{y}{3} \times -3) \geq -3 \times 3x => \boldsymbol{-3x > y \geq -9x}$.

Always be wary of changing the sign of inequality when multiplying or dividing it by a negative number.

The inequality can also be written as $\boldsymbol{3x < -y \leq 9x}$.

Similarly, if $3x > 6y - 3 > 6z$, we can divide the inequality by 3, so we get $\boldsymbol{x > 2y - 1 > 2z}$. Always remember that each part of the compound inequality should be divided.

3.13 Functions

We all know that area of a square equals square of its sides; $\boldsymbol{A = a^2}$.

It is clear that when we feed the value of the side of square, a, in the expression, we get the value of area, A. Here, '*a*' is called an **Independent variable** because '*a*' may have any value, it is not dependent on the value 'A'; whereas, 'A' is called a **Dependent variable** because 'A' cannot assume any value on its own, it is always dependent on the value '*a*'.

So, **Function** is a relationship between two variables-one of them is called is an **Independent variable** and the other one is called **Dependent variable**-such that for one value of independent variable, there is only one value of dependent variable.

In other words, '*a*' is an **Input**, and 'A' is an **Output**; the expression which can also be called a **Function**, expresses a relationship between the two variables.

For the formula $A = a^2$, we can write that $\boldsymbol{A = f(a)}$.

Similarly, if we write $\boldsymbol{y = f(x)}$, it means that $\boldsymbol{x}$ is an **independent variable**, and $\boldsymbol{y}$ is a **dependent variable**.

The significance of writing 'f' is to show the relationship between the two variables. It can also be written as $y = g(x)$ or $y = h(x)$. Here 'f', 'g', and 'h' are neither variables nor constants. You must NOT infer that in the expression $y = f(x)$, 'f' and 'x' are multiplied.

There would be only one independent variable and usually one dependent variable in quant questions.

Let us see couple of terminologies associated with function:

1. Domain:

Set of values an independent variable can take is called Domain. For the above function $A = f(a)$, the domain of 'a' is all positive real numbers.

2. Range:

Set of values a dependent variable can take corresponding to the values of independent variables is called Range. For the function $A = f(a)$, the range of 'A' is also all positive real numbers.

For a function $h(x) = |x + 4|$, the domain of x is **real number**, and the range of $h(x)$ is **positive real number**.

Example 1: If $f(x) = 2x^3 - 3x^2 - 2x + 4$, find $f(2)$.

Plug in the value of $x = 2$ in the function,

we get $f(2) = 2 \times 2^3 - 3 \times 2^2 - 2 \times 2 + 4 = 16 - 12 - 4 + 4 = 4$

Example 2: If $g(t) = 2t^3 - 3t^2 - \frac{3}{t^2} + \frac{2}{t^3}$, find $g(\frac{1}{t})$.

$g(1/t) = 2(1/t)^3 - 3(1/t)^2 - \frac{3}{(1/t)^2} + \frac{2}{(1/t)^3} = \frac{2}{t^3} - \frac{3}{t^2} - 3t^2 + 2t^3$; this is in fact $g(t)$.

So, $g\left(\frac{1}{t}\right) = g(t)$.

Compound functions:

When a function itself acts as an independent variable, it is called a compound function.
Thus, $f(g(x))$ is compound function, where $g(x)$ ploys the role of an independent variable for the function 'f'.

Example 3: If $f(x) = 2x + 2$, and $g(x) = x^2$, then find $f(g(x))$ & $g(f(x))$.

$f(g(x)) = f(x^2) = 2.x^2 + 2 = 2(x^2 + 1)$; replace the value of x in $f(x) = 2x + 2$ with x^2.

and $g(f(x)) = g(2x + 2) = (2x + 2)^2 = 4x^2 + 8x + 4$; replace the value of x in $g(x) = x^2$ with $(2x + 2)$.

Example 4: If $f(x) = 3x - 2$, and $f(g(x)) = 2x$, then find $g(x)$.

$f(g(x)) = 2x$

$=> 3 \times g(x) - 2 = 2x$; replace the value of x in $f(x) = 3x - 2$ with $g(x)$.

$=> 3 \times g(x) = 2x + 2$

$=> g(x) = \frac{2(x + 1)}{3}$

3.14 Geometry

3.15 Lines

A **line** is assumed to extend indefinitely in both directions. There is one and only one line between two distinct points.

A **line segment** is the part of a line between two points called end points. A line segment is denoted by its end points. It is denoted by $\overline{XY}$.

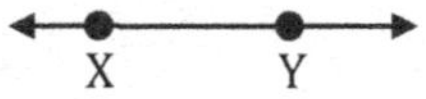

When a line segment is extended indefinitely in one direction, it is called a **ray**. A ray has one endpoint.

Pairs of lines: A pair of straight lines may be either:

- **Parallel:** Parallel lines are lines in the same plane that never intersect, even if extended infinitely.

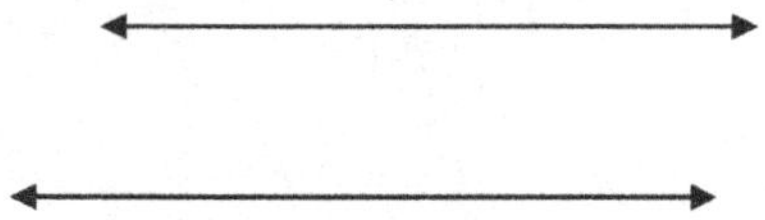

- **Intersecting:** A pair of lines in the same plane, if not parallel, must intersect at some point.

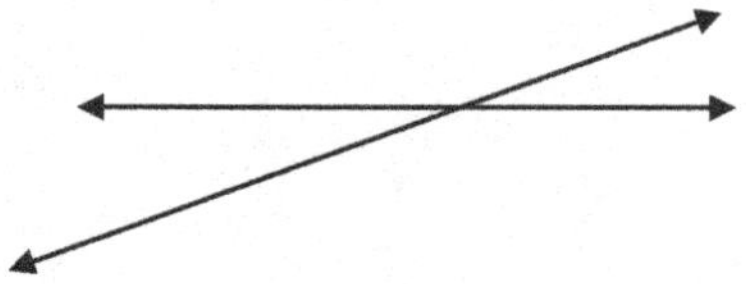

- **Perpendicular:** If the above lines intersect at right angles, then such lines are perpendicular to one another.

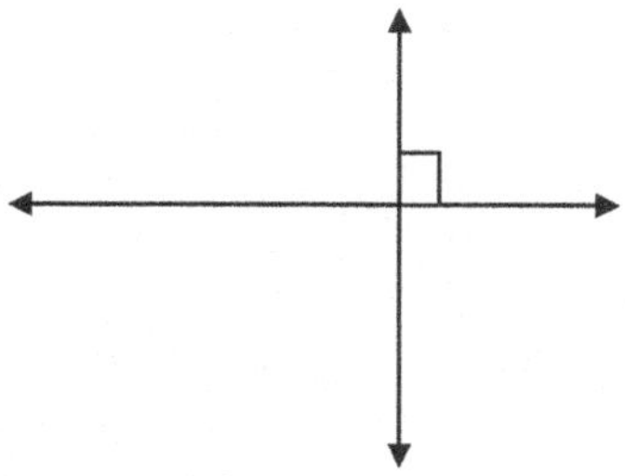

3.16 Angles

If two straight lines meet at a point they form an angle. The point is called the vertex of the angle and the lines are called the sides or rays of the angle.

Let us look at the important points associated with lines and angles:

1. An angle less than 90° is **acute**.	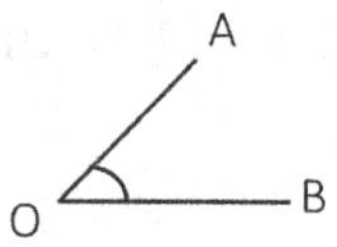
2. An angle equal to 90° is **right angle**.	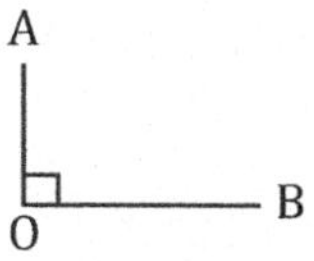
3. An angle greater than 90° but less than 180° is **obtuse**.	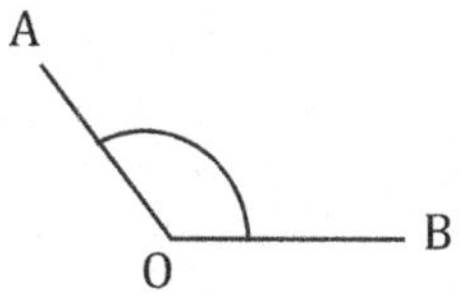
4. An angle of 180° is a **straight angle**.	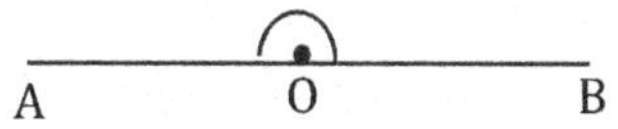
5. An angle greater than 180° but less than 360° is **reflex.**	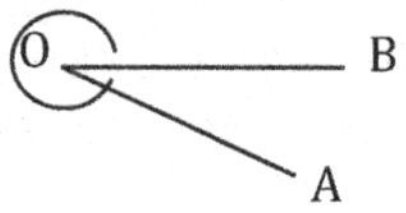

6. Two angles are **adjacent** if they have the same vertex with a common side and one angle is not inside the other. • ∠BOC and ∠COD are adjacent • ∠COD and ∠EOD are not adjacent	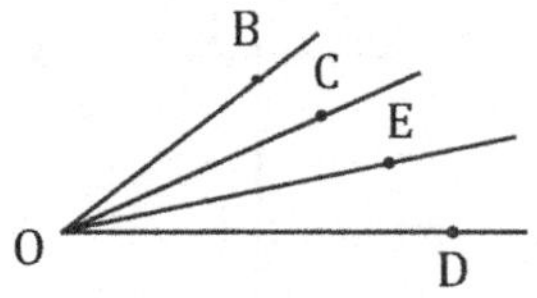
7. If the sum of two adjacent angles is 180°, then they are **supplementary**. • ∠AOC and ∠DOC are supplementary.	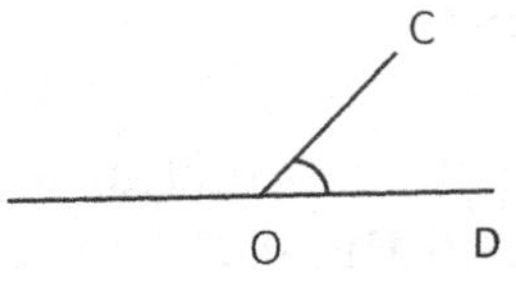
8. If the sum of two adjacent angles is 90°, then they are **complementary**. • ∠BOC and ∠DOC are complementary.	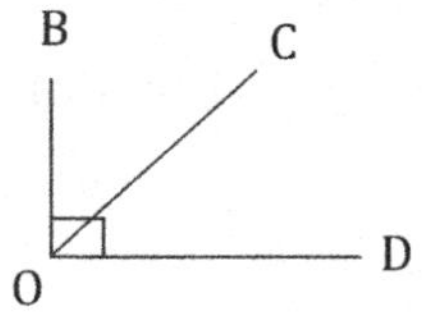
9. **Vertically Opposite Angles:** When two lines intersect, two pairs of vertically opposite angles are formed. Vertically opposite angles are equal. • Here, ∠C and ∠A are equal. • Similarly, ∠D and ∠B are equal.	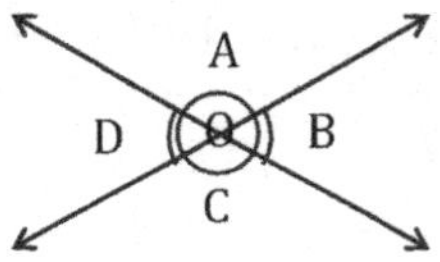

10. **Alternate and Corresponding Angles:** In the diagram below, AB and UV are parallel lines with DE as a transversal. (Note: A transversal refers to any straight line that intersects a pair of straight line.) This creates pairs of corresponding and alternate angle as shown below. • Alternate angles are equal $\angle Q = \angle S$ and $\angle P = \angle R$ • Interior angles on same side are supplementary $\angle P + \angle S = \angle Q + \angle R = \angle 180^\circ$ • Corresponding angles are equal $\angle N = \angle S$; $\angle M = \angle R$; $\angle P = \angle T$; $\angle Q = \angle U$	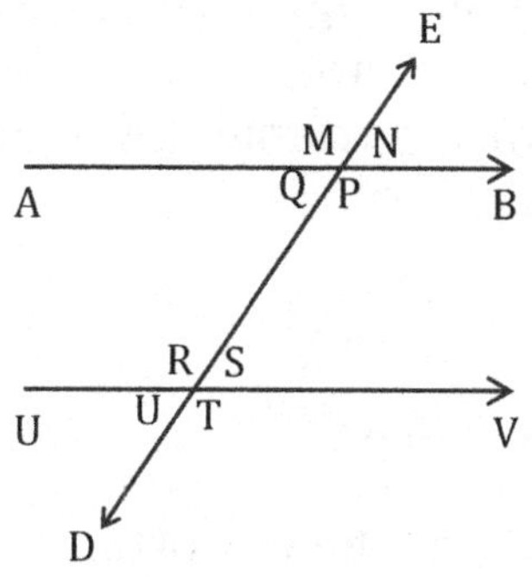

3.17 Triangles

A triangle is a figure bounded by three line segments in a plane. In the diagram, the line segments are AB, BC and CA, also called its sides; and A, B and C are called the vertices of the triangle. A triangle can also be called a 3-sided polygon.

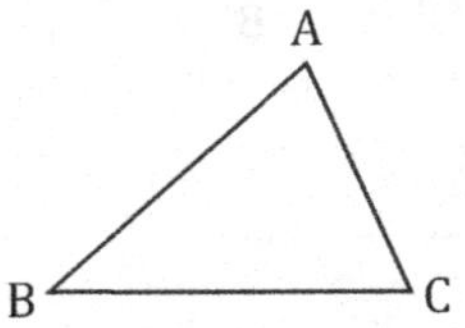

Let us look at the important points associated with triangles

1. Sum of any two sides in a triangle is greater than the third side. Also, any side is less than half of the perimeter of the triangle. Note: What happens, if the sum of two sides is NOT greater than the third side? Let us observe using the diagram below. Let AB be one side of the triangle. Let the third vertex be C. To determine the position of C, we draw arc with radius AC and BC with centres at A and B respectively. The point of intersection of the arc is the position of C. If $AC + BC \not> AB$, then the arc never intersects as shown below. Thus, there is no triangle at all! 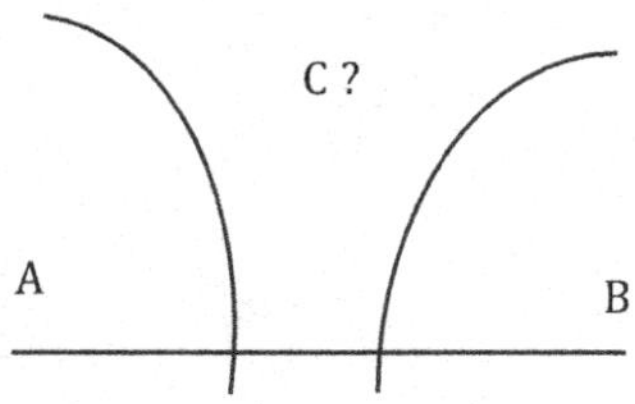 	Thus, in triangle ABC, we simultaneously have $AB + BC > AC \ldots$ (i) $AB + AC > BC \ldots$ (ii) $AC + BC > AB \ldots$ (iii) Let us choose any of the three inequalities above, say (i): adding AC to both sides $AB + BC + AC > AC + AC$ $\Rightarrow 2AC <$ Perimeter $\Rightarrow AC < \frac{1}{2} \times$ Perimeter Similarly, the inequalities for the other two sides follow.

2. The greatest side in a triangle has the greatest angle opposite to it and vice-versa. Similarly, the smallest side in a triangle has the smallest angle opposite to it and vice-versa.	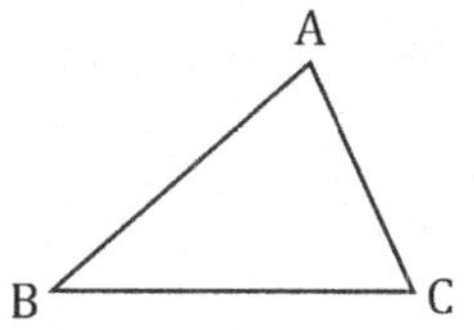 AB > BC > AC => ∠C > ∠A > ∠B
3. Sum of the interior angles is 180°. Sum of its exterior angles is 360°. In the figure alongside, the interior angles are i_a, i_b and i_c and the exterior angles are e_a, e_b and e_c. Note: Have you ever wondered: Why is the sum of the interior angles of a triangle equal to 180°? (Find out - At the end of this chapter)	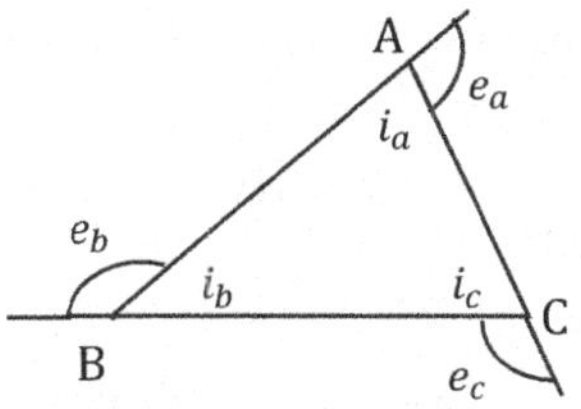 Thus, we have $i_a + i_b + i_c = 180°$ $e_a + e_b + e_c$ $= 180° - i_a + 180° - i_b + 180° - i_c$ $= 180° \times 3 - (i_a + i_b + i_c)$ $= 540° - 180°$ $= 360°$
4. Any exterior angle in a triangle is equal to sum of the opposite interior angles.	This follows from the above rule #3. Let us see how Since $i_a + i_b + i_c = 180°$, we have $i_a + i_b = 180° - i_c$ $=> i_a + i_b = e_c$

5. A triangle with no two sides equal is called a **scalene** triangle.	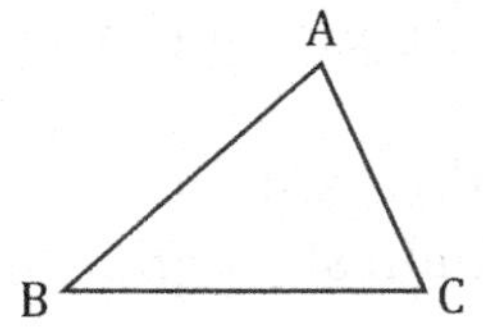 Here, AB ≠ BC ≠ CA, and ∠C ≠ ∠A ≠ ∠B
6. If EXACTLY two sides of a triangle are equal, it is called an **isosceles** triangle. The angles opposite to the equal sides are also equal. **Note:** In an isosceles triangle, a perpendicular drawn from the vertex to the 'UNEQUAL' side bisects the side.	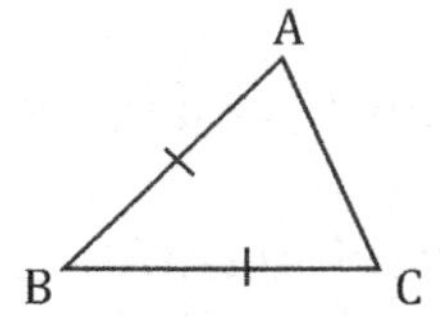 Here, AB = BC ≠ AC => ∠C =∠A ≠ ∠B 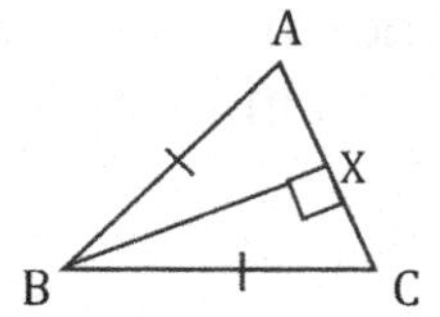 Here, BX ⊥ AC => AX = CX, and ∠ABX = ∠CBX
7. If all the three sides of a triangle are equal, it is called an **equilateral** triangle. Each angle is equal to 60°. • Height of an equilateral triangle $= \frac{\sqrt{3}}{2} \times$ (Length of its side) • Area of an equilateral triangle $= \frac{\sqrt{3}}{4} \times$ (Length of its side)^2 • Perimeter of an equilateral triangle = 3 × Length of its side	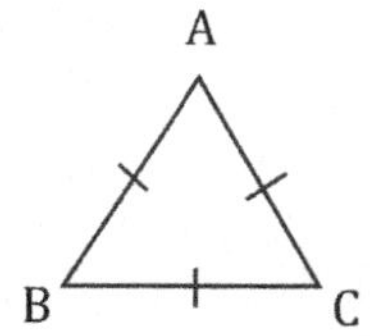 Here, AB = BC = CA ∠C = ∠A = ∠B = 60°

8. If ONE angle of a triangle is 90°, it is called a **right-angled** triangle.

   If ABC is right-angled at A, from Pythagoras' theorem we have

   $\boldsymbol{AB^2 + AC^2 = BC^2}$

   A few of the integer set of values of the sides AB, AC and BC satisfying Pythagoras' theorem are:

   - **3, 4, 5:** $3^2 + 4^2 = 5^2$
   - **6, 8, 10:** $6^2 + 8^2 = 10^2$
   - **5, 12, 13:** $5^2 + 12^2 = 13^2$
   - **7, 24, 25:** $7^2 + 24^2 = 25^2$
   - **8, 15, 17:** $8^2 + 15^2 = 17^2$

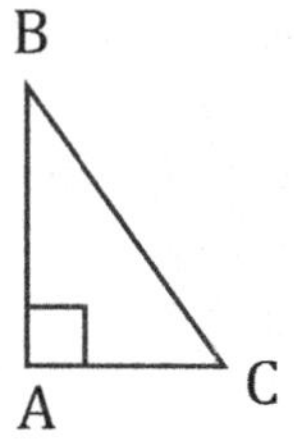

Another way of looking at Pythagoras' theorem:

The area of the square constructed on BC as its side equals the sum of areas of the squares constructed on AB and AC as the sides.

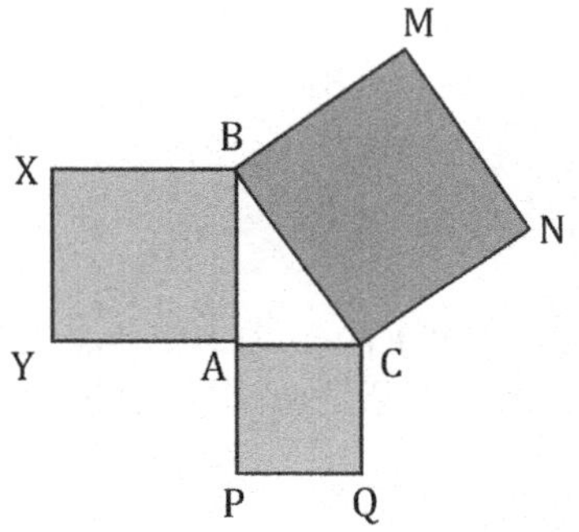

Thus, we have

Area BMNC
= Area ACQP + Area ABXY

9. If ALL the angles of a triangle are less than 90°, it is called an **acute-angled** triangle.

   Since ∠A, ∠B and ∠C are all acute, we have

   - $AB^2 + AC^2 > BC^2$
   - $AB^2 + BC^2 > AC^2$
   - $AC^2 + BC^2 > AB^2$

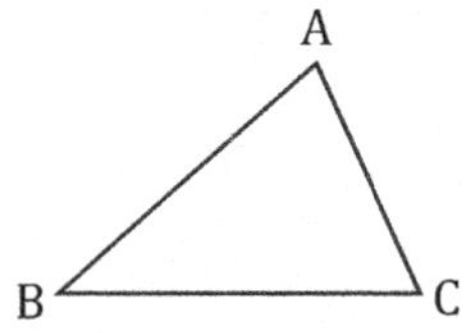

Here, $\angle C < 90°$, $\angle A < 90°$ and $\angle B < 90°$

<table>
<tr><td>

10. If ONE angle of a triangle is more than 90°, it is called an **obtuse-angled** triangle.

Since C is obtuse, we have

- $BC^2 + AC^2 < AB^2$

</td><td>

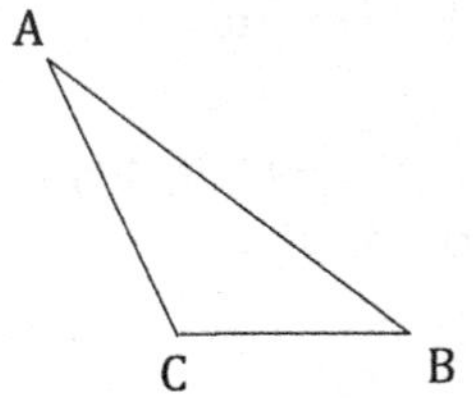

Here, $\angle C > 90^\circ$, $\angle A < 90^\circ$ and $\angle B < 90^\circ$

</td></tr>
<tr><td>

11. **Area** of a triangle:

- Area of $\triangle ABC = \frac{1}{2} \times$ Base $\times$ Height

$= \frac{1}{2} \times BC \times AD$

$= \frac{1}{2} \times AC \times BE$

$= \frac{1}{2} \times AB \times CF$

Thus, from the above, we have
$BC \times AD = AC \times BE = AB \times CF$

Thus, we have

- $AD : BE : CF = \frac{1}{BC} : \frac{1}{AC} : \frac{1}{AB}$

Thus, the ratio of the altitudes equals the ratio of the reciprocals of the corresponding sides of a triangle.

</td><td>

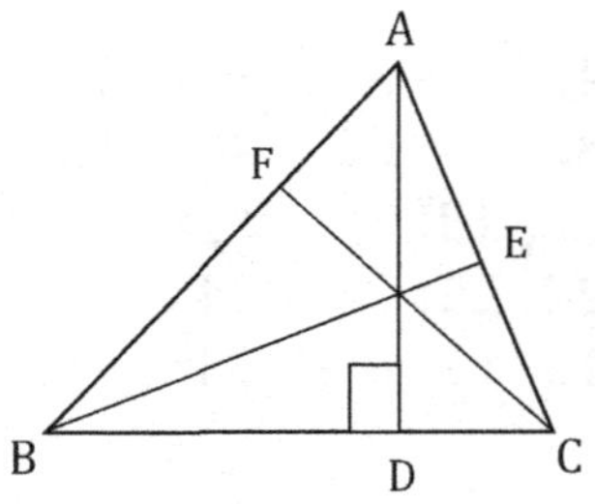

</td></tr>
</table>

12. For triangles lying between the same set of parallel lines (l and m), the ratio of the areas of the triangles equals the ratio of the corresponding bases of the triangles. Since the triangles lie between the same set of parallel lines, their heights are the same, and hence, the ratio of their area equals the ratio of their bases.	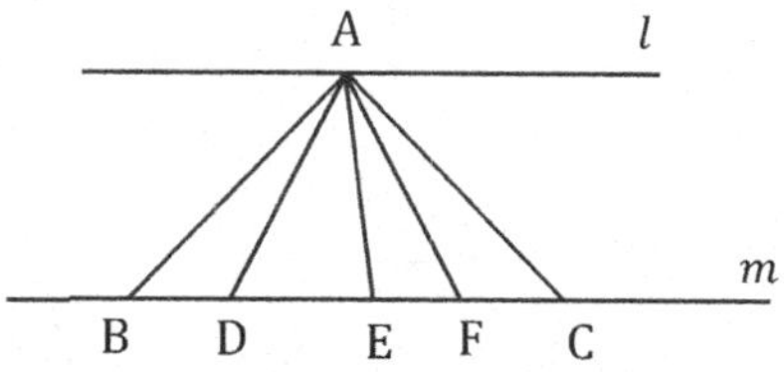 Here, lines from the vertex A are drawn to the base BC intersecting at D, E and F. Thus, triangles ABD, ADE, AEF and AFC lie between the same parallel lines (hence, they have the same height) => Area of △ABD : Area of △ADE : Area of △AEF : Area of △AFC = BD : DE : EF : FC

Let us now look into relations between different triangles. There are two important areas to focus on: Congruency and Similarity. Let us discuss them one by one.

A. **Congruent Triangles:** Two triangles are congruent if they are identical to each other in all aspects. Thus, the corresponding sides and angles of the triangles are equal.

There are four ways to prove congruency:

(a) **SAS** (Side-Angle-Side)

Triangle ABC ≅ Triangle XYZ if AB = XY; ∠A = ∠X; AC = XZ (Here the angle is the included angle)

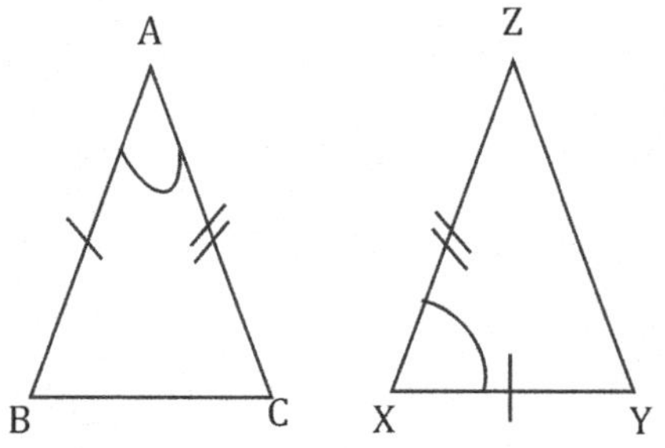

Note: Triangles MNP and DEF are NOT congruent:

Since: MN = DF; ∠N = ∠E; MP = DE
(∠N & ∠E are not the included angles)

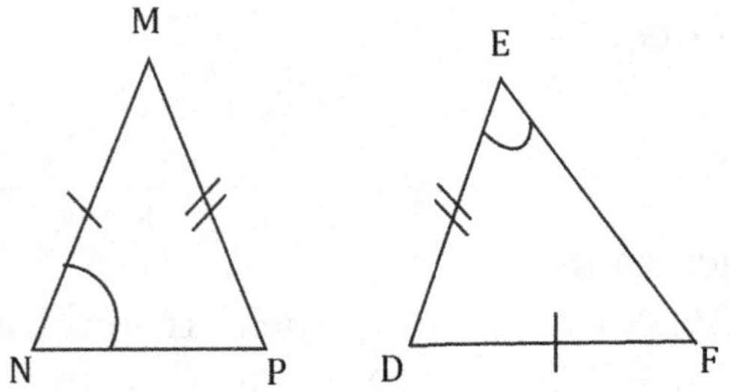

(b) **AAS** (Angle-Angle-Side)

Triangle ABC ≅ Triangle XYZ if ∠B = ∠Y; ∠C = ∠Z; AC = XZ.
Here, however, the side NEED NOT be the included side
(If two angles are equal, the third angle will also be equal)

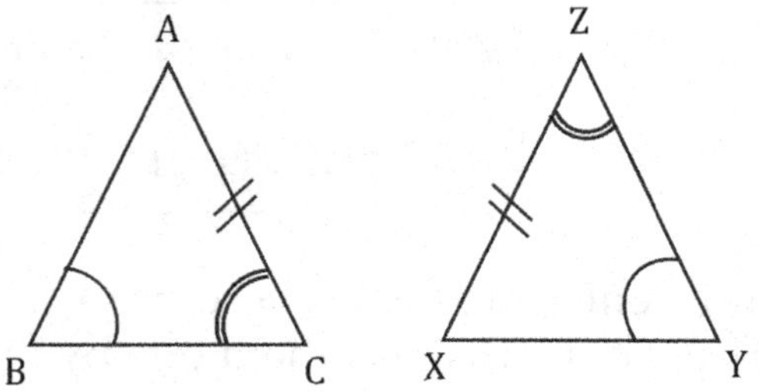

(c) **SSS** (Side-Side-Side)

Triangle ABC ≅ Triangle XYZ if AB = XY; AC = XZ; BC = YZ

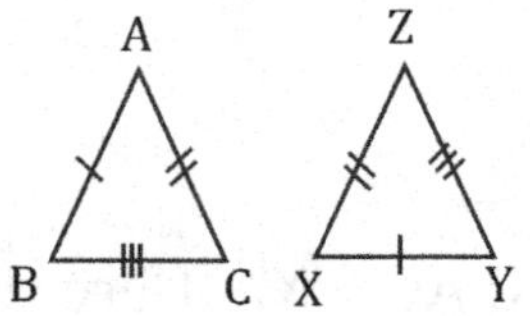

(d) **RHS** (Right angle-Hypotenuse-Side)

Triangle ABC ≅Triangle XYZ if ∠B = ∠Y = 90°;
AC = XZ (hypotenuse); AB = XY.

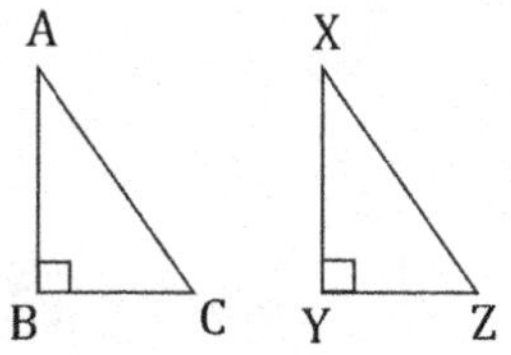

B. Similar Triangles: Similar triangles are similar in shape, need not have same size but are proportional i.e. the ratio of their sides is constant.

Two or more triangles can be called similar if:

(a) **AAA** (Angle-Angle-Angle)

Triangle ABC ~ Triangle XYZ if $\angle B = \angle Y$; $\angle C = \angle Z$
Note: Since $\angle A + \angle B + \angle C = \angle X + \angle Y + \angle Z = 180°$
$\Rightarrow \angle A = \angle X$

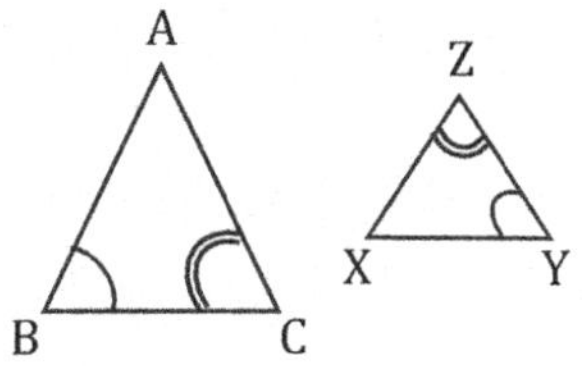

(b) **SAS** (Side ratio-Angle-Side ratio)

Triangle ABC ~ Triangle XYZ if $\angle A = \angle X$; $\frac{AB}{XY} = \frac{AC}{XZ}$
(Here the angle is the included angle)

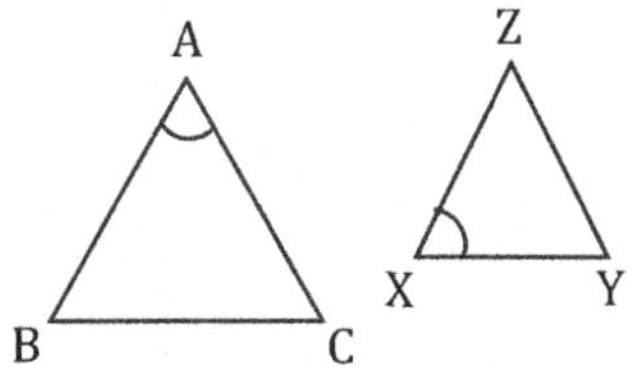

Let us look at three important results for similar triangles:

If two triangles are similar

- Ratio of their corresponding sides equals the ratio of any corresponding length measure of the triangles; example: ratio of their altitudes or ratio of their medians, etc.
- Ratio of the areas of the triangles equals the SQUARE of the ratio of their corresponding sides:

  For any two similar triangles ABC and XYZ, we have

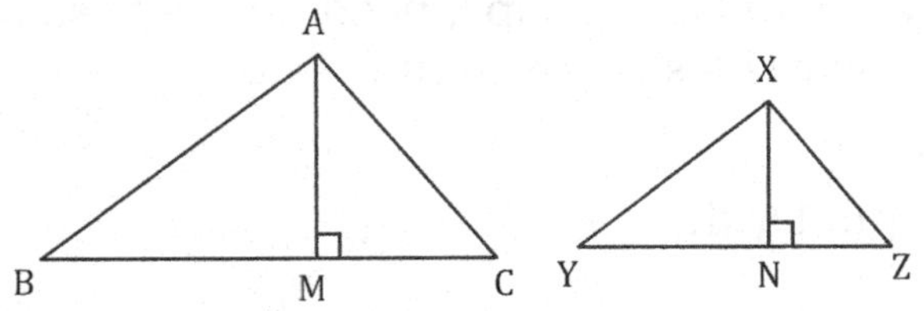

Let the ratio of sides of the two triangles be k.

Thus, we have

$$\frac{AB}{XY} = \frac{BC}{YZ} = \frac{CA}{ZX} = \frac{AM}{XN} = k$$

$$\frac{(\text{Area of } \triangle \text{ ABC})}{(\text{Area of } \triangle \text{ XYZ})} = \left(\frac{\frac{1}{2} \times BC \times AM}{\frac{1}{2} \times YZ \times XN}\right) = \left(\frac{BC \times AM}{YZ \times XN}\right) = \left(\frac{BC}{YZ}\right) \times \left(\frac{AM}{XN}\right) = k \times k = k^2$$

- Ratio of their perimeters of the triangles equals the ratio of their corresponding sides.

A few important rules are now presented below. Effective use of these rules can reduce the time taken to solve a problem.

(1) **Basic Proportionality Theorem:** A line parallel to one side of a triangle divides the other two sides proportionally i.e. it creates two similar triangles, one inside the other, sharing a common vertex.

In triangle ABC, DE is drawn parallel to BC.

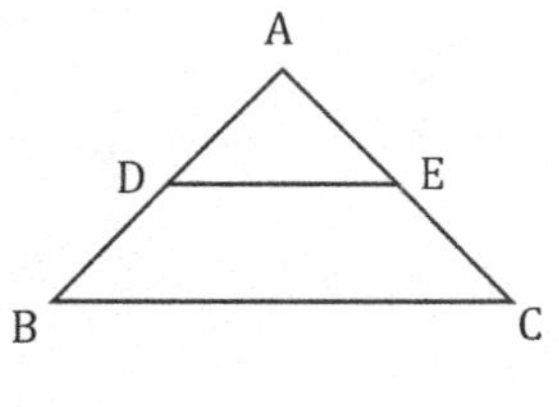

Thus, we have $\triangle ABC \sim \triangle ADE$.

$$\Rightarrow \frac{AD}{AB} = \frac{AE}{AC} \Rightarrow \frac{AD}{DB} = \frac{AE}{EC}$$

(2) **Mid-point Theorem:** The straight line joining the mid-points of two sides of a triangle is parallel to the third side and equal to half the third side.

(3) **Angle Bisector Theorem:** A bisector of any angle of a triangle will divide the opposite side in the same proportion as the two sides adjacent to the angle.

In $\triangle ABC$, AP is the angle bisector of $\angle A$. Thus, we have

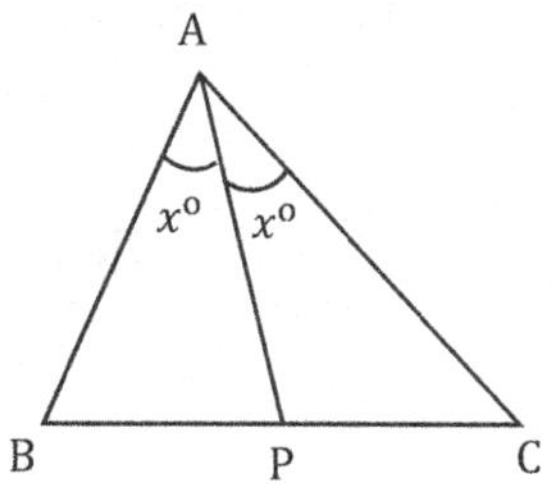

$$\frac{AB}{AC} = \frac{BP}{PC}$$

(4) **Right Triangles:**

- In any right triangle, the length of the median to the hypotenuse equals half the length of the hypotenuse.

  Here, AX is the median drawn to the hypotenuse BC.

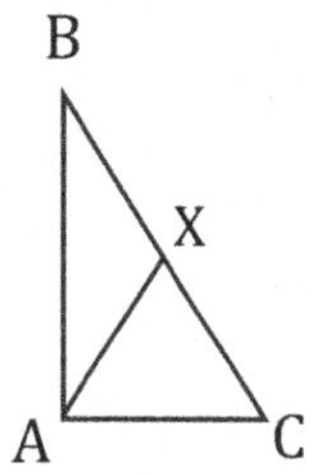

  Thus, we have

  AX = BX = CX

- If AX is drawn ⊥ to the hypotenuse intersecting the hypotenuse BC at X, we have three similar triangles (AXB, CXA and CAB). Thus, we have the following relations:

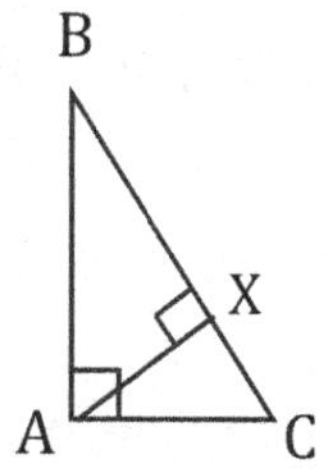

  - $AX^2 = BX \times CX$ (from similarity of triangles AXB and CXA)
  - $AB^2 = BX \times BC$ (from similarity of triangles AXB and CAB)
  - $AC^2 = CX \times CB$ (from similarity of triangles CAX and CAB)
  - $AB \times AC = AX \times BC$ (equating area: $\frac{1}{2} \times AB \times AC = \frac{1}{2} \times AX \times BC$)

- $30^\circ - 60^\circ - 90^\circ$ Triangle: Basically, a $30^\circ - 60^\circ - 90^\circ$ triangle is half of an equilateral triangle.

  It can be seen that half of triangle XBC is the $30^\circ - 60^\circ - 90^\circ$ triangle ABC.

  The ratio of the sides of a $30^\circ - 60^\circ - 90^\circ$ triangle is shown:

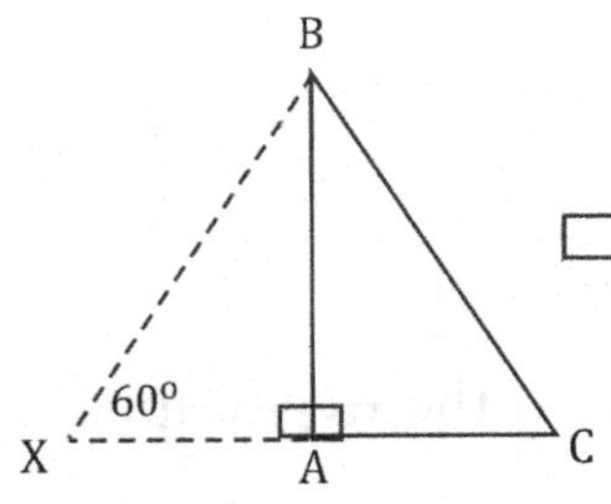

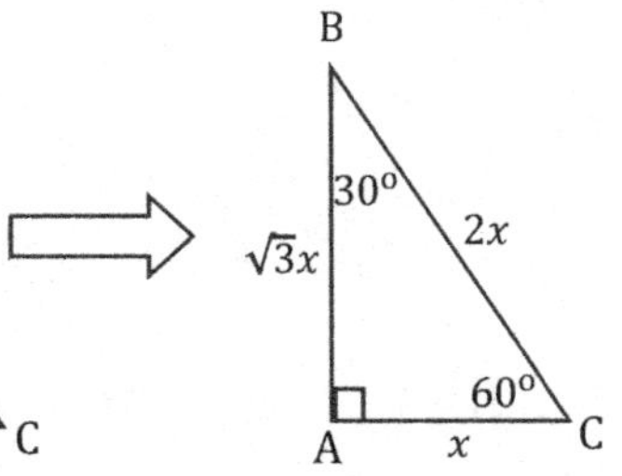

- $45^\circ - 45^\circ - 90^\circ$ Triangle: Basically, a $45^\circ - 45^\circ - 90^\circ$ triangle is half of a square. It can be seen that half of square ABXC is the $45^\circ - 45^\circ - 90^\circ$ triangle ABC.

  The ratio of the sides of a $45^\circ - 45^\circ - 90^\circ$ triangle is shown:

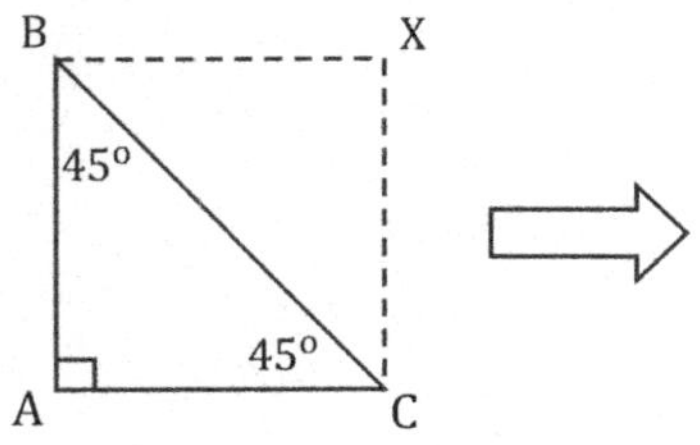

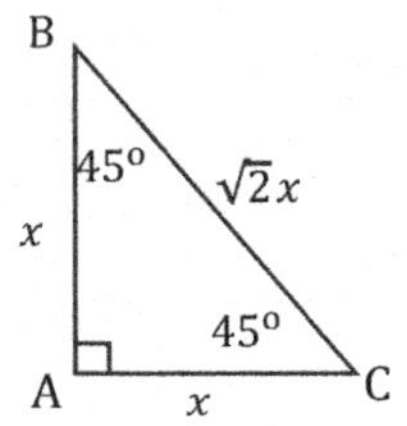

(5) **Maximum / minimum area and perimeter of triangles:**

- Among all triangles of the same area, the equilateral triangle has the minimum perimeter.
- Among all triangles of the same perimeter, the equilateral triangle has the maximum area.
- Among all isosceles triangles of the same area, the right-angled isosceles triangle has the minimum perimeter.
- Among all isosceles triangles of the same perimeter, the right-angled isosceles triangle has the maximum area.
- Among all triangles given the lengths of two sides, the triangle having those two sides perpendicular to one another has the maximum area.

Why is the sum of the interior angles of a triangle 180°?

In the triangle ABC, let us draw a line XY through A, parallel to BC.

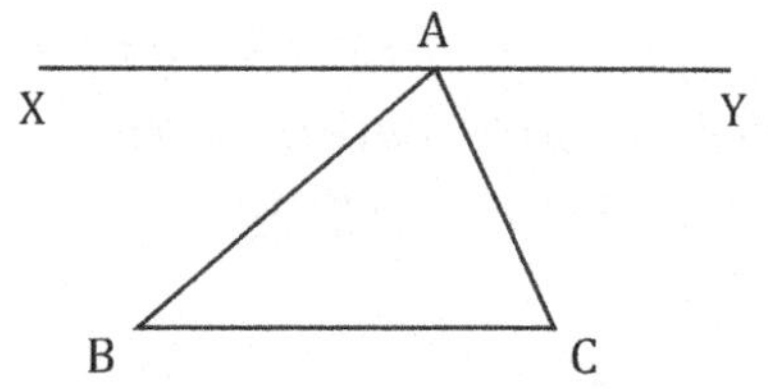

Thus, we have

$\angle ABC = \angle XAB$, and $\angle ACB = \angle YAC$

Since XY is a straight line, we have

$\angle XAB + \angle BAC + \angle YAC = 180^\circ$

$\Rightarrow \angle ABC + \angle BAC + \angle ACB = 180^\circ$

3.18 Polygons

A polygon is a closed figure bounded by line segments. Some common names of polygons:

A polygon with:

3 sides = **triangle**	6 sides = **hexagon**	9 sides = **nonagon**
4 sides = **quadrilateral**	7 sides = **heptagon**	10 sides = **decagon**
5 sides = **pentagon**	8 sides = **octagon**	Infinite sides = **circle**

Nomenclature:

- A polygon with all sides equal AND all angles equal is called a **regular polygon**.
- A polygon with no interior angle greater than 180° is called a **convex polygon**.
- A polygon with at least one interior angle greater than 180° is called a **concave polygon**.

Let us look at the important points associated with polygons: If the number of sides of a polygon be n, we have

<table>
<tr>
<td>1. Sum of all interior angles
$= (n-2) \times 180^\circ$</td>
<td>Any polygon of n sides can be broken down to $(n-2)$ non-overlapping triangles as shown:
<table>
<tr><td>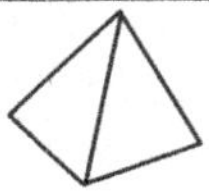</td><td>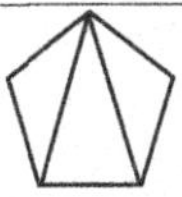</td><td>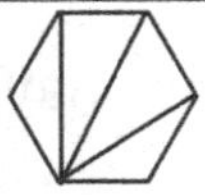</td></tr>
<tr><td>$n = 4$
of $\triangle$s
$= 4-2 = 2$</td><td>$n = 5$
of $\triangle$s
$= 5-2 = 3$</td><td>$n = 6$
of $\triangle$s
$= 6-2 = 4$</td></tr>
</table>
Thus, sum of interior angles is simply the sum of the angles of all $(n-2)$ triangles

$= (n-2) \times 180^\circ$</td>
</tr>
<tr>
<td>2. Sum of all exterior angles
$= 360^\circ$</td>
<td>Sum of all exterior angles

= (Sum of all interior & exterior angles) – (Sum of interior angles)

$= n \times 180^\circ - (n-2) \times 180^\circ$

$= 360^\circ$</td>
</tr>
<tr>
<td>3. Number of diagonals
$= \dfrac{n(n-3)}{2}$</td>
<td>Number of diagonals

= (# of ways in which any two vertices of the polygon can be joined) – (# of sides)

$= C_2^n - n = \dfrac{n(n-1)}{2} - n$
$= \dfrac{n(n-3)}{2}$</td>
</tr>
</table>

4. Each interior angle of a regular polygon $= \frac{(n-2) \times 180^\circ}{n}$	Each interior angle of a regular polygon $= \frac{\text{Sum of all interior angles}}{\text{\# of angles}}$ $= \frac{(n-2) \times 180^\circ}{n}$
5. Each exterior angle of a regular polygon $= \frac{360^\circ}{n}$	Each exterior angle of a regular polygon $= \frac{\text{Sum of all exterior angles}}{\text{\# of angles}}$ $= \frac{360^\circ}{n}$
6. Area of a regular polygon $= \frac{1}{2} \times (\text{perimeter}) \times (\perp \text{ from centre})$	Area of a regular polygon is simply the sum of the areas of n triangles formed by joining the vertices to the centre of the polygon as shown: 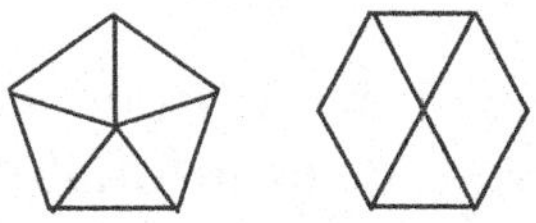 Thus, area of the polygon $= n \times \frac{1}{2} \times (\text{Side}) \times (\perp \text{ from centre})$ $= \frac{1}{2} \times (n \times \text{Side}) \times (\perp \text{ from centre})$ $= \frac{1}{2} \times (\text{perimeter}) \times (\perp \text{ from centre})$

3.19 Quadrilateral

A quadrilateral has four sides and sum of all four angles is 360°.

Let us look at the properties of the main quadrilaterals:

(1) **Parallelogram:** The properties are given below:

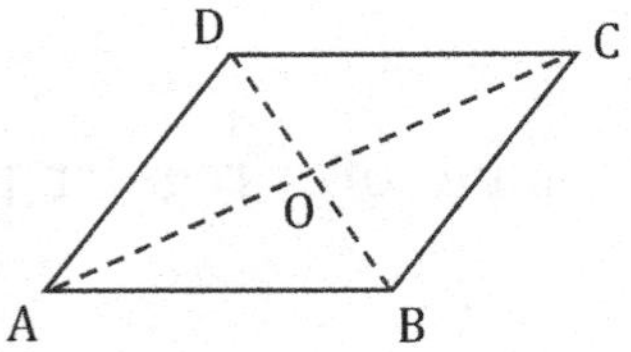

- Opposite sides are parallel and equal: $AB \parallel CD$, $AB = CD$, $AD \parallel BC$, $AD = BC$
- Opposite angles are equal: $\angle A = \angle C$ & $\angle B = \angle D$
- Diagonals bisect each other: $AO = OC$ & $BO = DO$
- Sum of any two adjacent angles = 180°: $\angle A + \angle D = 180^\circ$ & $\angle C + \angle B = 180^\circ$
- Straight lines joining the midpoints of adjacent sides of any quadrilateral form a parallelogram
- Area = Base $\times$ Height

(2) **Rectangle:** All the above properties of a parallelogram hold true for a rectangle. However, there are few additional properties:

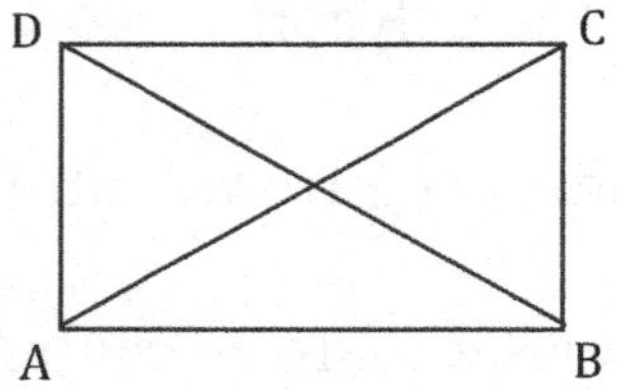

- All angles are equal and equal to a right angle: $\angle A = \angle B = \angle C = \angle D = 90^\circ$
- Diagonals are equal: $AC = BD$
- Perimeter = $2 \times$ (Length + Breadth)
- Area = Length $\times$ Width
- Length of Diagonal = $\sqrt{\text{Length}^2 + \text{Breadth}^2}$

(3) **Square:** All the properties mentioned above for all the figures hold true for a square.

Thus, we have

- All sides are equal: AB = BC = CD = DA
- All angles are right angles.
- The diagonals are equal and bisect each other at right angles.
- Perimeter of a square = $4a$, where a is the length of a side.
- Length of a diagonal = $a\sqrt{2}$
- Area = $(\text{Side})^2 = \frac{1}{2} \times (\text{Diagonal})^2 = a^2$

Maximum / minimum area of quadrilaterals:

- Among all quadrilaterals with the same area, the square has the least perimeter.
- Among all quadrilaterals with the same perimeter, the square has the greatest area.

3.20 Circles

Let us look at the important points associated with circles

1. Perpendicular from the centre bisects the chord i.e. OZ bisects XY 2. Similarly, if Z is the mid-point of XY, then OZ is perpendicular to XY	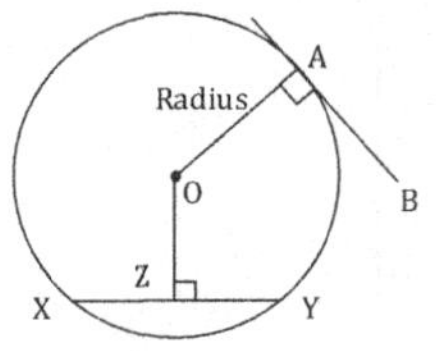 O: Centre of the circle OA: Radius XY: Chord
3. Angle subtended at the centre by an arc (QR) is equal to twice the angle made by the same arc at any other point on the remaining circumference => $\angle QOR = 2 \times \angle QPR$	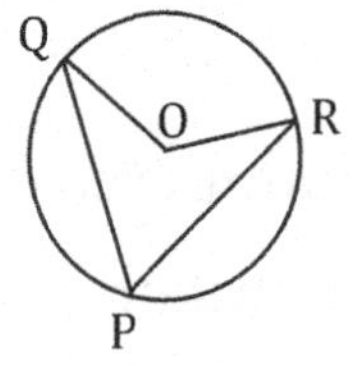
4. The angle between a tangent (QP) and a chord (RP) at the point of contact (P) is equal to the angle in the alternate segment => $\angle QPR = \angle PSR$ 5. As a result of the above, triangles PRQ and SPQ are similar, which gives us: $PQ^2 = SQ \times RQ$	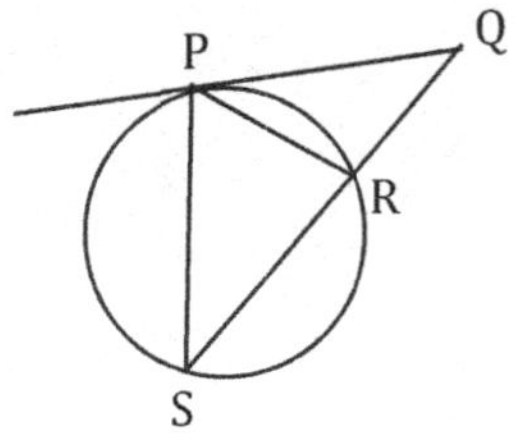

6. If two chords intersect: $AE \times BE = CE \times DE$	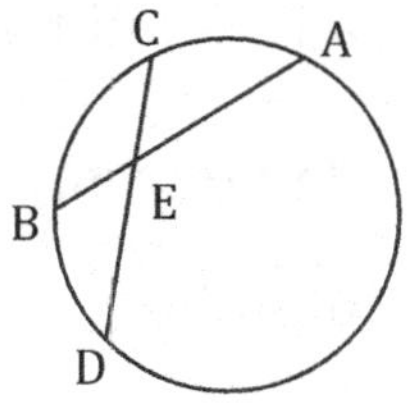
7. Area of a circle of radius $r = \pi r^2$ 8. Circumference of the above circle $= 2\pi r$ 9. Length of the arc subtending angle θ at the centre of a circle of radius r $= \left(\frac{\theta}{360}\right) \times 2\pi r$ 10. Area of the sector formed by the above arc $= \left(\frac{\theta}{360}\right) \times \pi r^2$ $= \frac{1}{2} \times r \times \left\{\left(\frac{\theta}{360}\right) \times 2\pi r\right\}$ $= \frac{1}{2} \times r \times$ (Length of arc) 11. Perimeter of the sector above = (Length of the arc) + (2 × Radius) $= \left(\frac{\theta}{360}\right) \times 2\pi r + 2r$	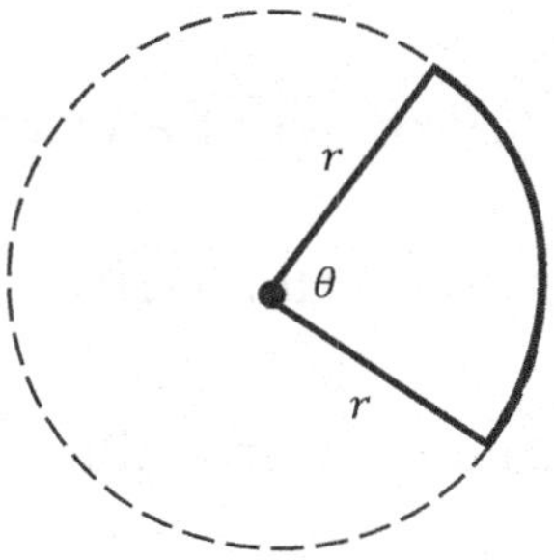

Maximum / minimum area of any polygon:

- Among all polygons with the same perimeter, the circle has the maximum area.
- Among all polygons with the same area, the circle has the least perimeter.

3.21 3-D Geometry

We will consider a few solid shapes.

Few important 3-D figures:

(1) **Cuboid:** A cuboid is a prism having six rectangular faces.

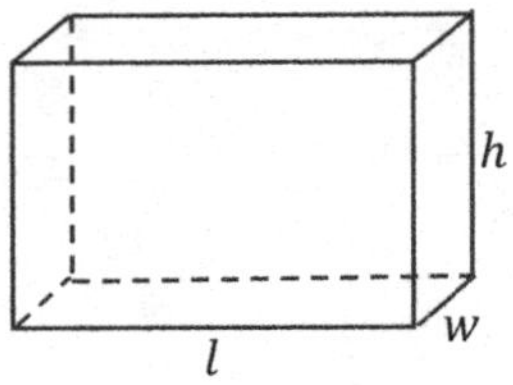

- **Volume** = $l \times w \times h$, where l = length, w = width, h = height

  $= \sqrt{\text{Product of areas of the mutually perpendicular faces}}$

  $= \sqrt{(lw)(wh)(lh)} = lwh$

- **Area of four walls** = $2(l + w) \times h$

- **Total surface area** = $2(lw + wh + lh)$

- **Body Diagonal** = $\sqrt{(l^2 + w^2 + h^2)}$

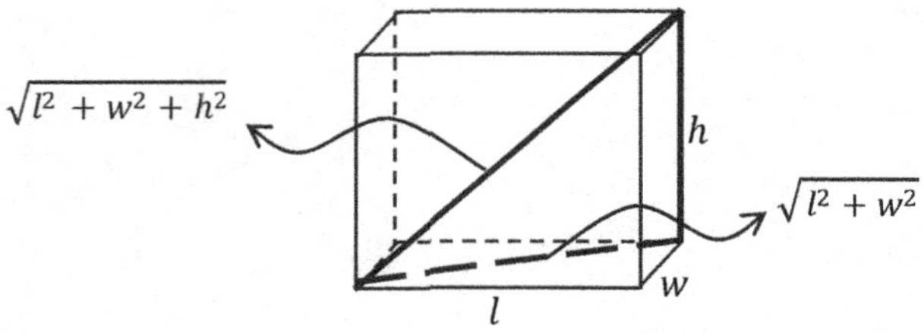

The 'Body Diagonal' is the longest line joining any two vertices of the cuboid.

(2) **Cube:** A prism whose every face is a square is called a cube.

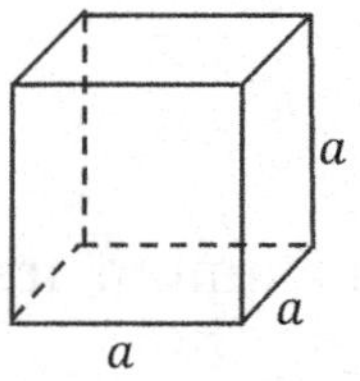

- **Volume** = a^3,where a = edge of the cube
- **Total surface area of the cube** = Area of 6 squares of area $a^2 = 6a^2$

- **Body Diagonal of a cube** = $a\sqrt{3}$

(3) **Cylinder:** If a rectangle is revolved about one of its sides as the axis, the solid thus formed is a right circular cylinder.

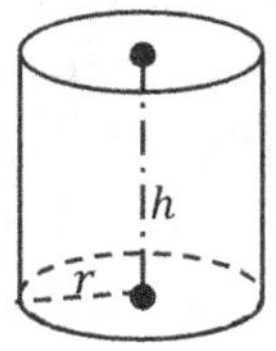

- **Volume** = $\pi r^2 h$, where r = radius of base, h = height
- **Curved surface area** = $2\pi rh$
- **Total surface area** = $2\pi r\,(r + h)$
- **Longest distance between two points** = $\sqrt{(2r)^2 + h^2}$

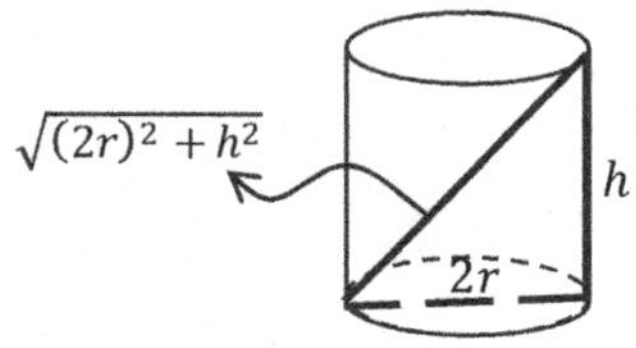

(4) **Cone:** If a right triangle is revolved about one of its sides containing the right angle as the axis the solid formed is called a cone.

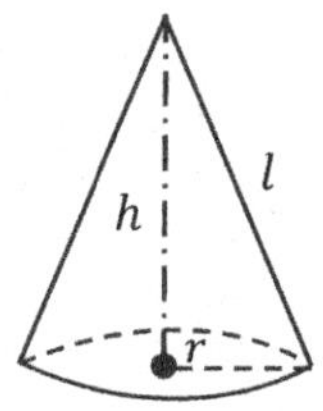

- **Volume** = $\frac{1}{3}\pi r^2 h$, where r = radius of base, h = height
- **Curved surface area** = πrl where l = slant height = $\sqrt{(r^2 + h^2)}$
- **Total surface area** = $\pi r(r + l)$

(5) **Sphere:** When a circle is revolved about its diameter, the solid thus formed is called a sphere.

- **Volume** = $\frac{4}{3}\pi r^3$, where r = radius of the sphere
- **Surface area** = $4\pi r^2$

Note: The test seldom asks you questions applying the formulae for volume or surface area of the sphere, cone or pyramids in general. However, you should remember that the volume of a cone is proportional to r^2h and that for the sphere is proportional to r^3.

The next section of the book gives you over 100+ practice equations. It does not have questions that the DSAT specify asks. The purpose of the book is to sharpen your math fundamentals. Once you sharpen your fundamentals on the SAT math, you will be in a better position to solve SAT math questions.

Additionally, the DSAT math book has questions that the SAT asks.

Chapter 4

Practice Questions

4.1 Numbers and Digits

Q1. Identify rational and irrational numbers.

(1) $\pi = 3.141592653589793238...$

(2) 12.2367324902...

(3) 1.7373737373...

(4) $0.\overline{02}$

(5) π^2

Q2. Write $6.\overline{52}$ in a rational fraction.

Q3. Write $0.\overline{524}$ in a rational fraction.

Q4. There are two positive integers a and b, and two negative integers c and d. Each ends with digit '2'. What would be the units digits of $(a + b + c + d)$?

Q5. If a decimal number $25.4n8$, rounded to the nearest tenth is less than 25.5, what is the set of values for n?

Q6. What is the value of $77,567,893 \times 10^{-6}$ rounded off to the nearest whole number?

4.2 Even/Odd and Consecutive Integers

Q1. If $a + b$ is even, is ab even? Given that a and b are positive integers.

Q2. If (pq) and r are even, is $(p + r)$ even?

Q3. If x, y, and z are consecutive integers, is xyz even?

Q4. If x is even, y, and z are odd, is $z(x + y)$ even?

Q5. What if the question is: If z is even and x and y are integers, is $z(x + y)$ even?

Q6. What if the question is: If y & z are odd, and x is an integer, is $z(x + y)$ even?

Q7. What if the question is: If y is even, z is odd, and x is an integer, is $z(x + y)$ even?

Q8. If n is an even integer, which of the following is an odd integer?

A. $n^2 + 3$

B. $3n^2 - 3$

C. $n^3 + 2$

D. $2n + 3$

E. $-3n + 3$

F. $n(n^4 - 3n^3 + 14n^2 - 4)$

G. $\sqrt{n^2 + 12}$

Q9. If p, q, r, s and t are consecutive integers, is their sum even?

Q10. What if the questions is: If $p, q, r,$ and s are consecutive integers, is their sum even?

4.3 Divisibility

Q1. The remainder is 37 when a number is divided by 10,000. What is the remainder when the same number is divided by 1,000?

Q2. If a is divisible by 7 and 18, is $\frac{a}{42}$ an integer?

Q3. If 70 is a factor of x, is 20 a factor of x?

Q4. If p is divisible by 15 and 14, is it divisible by 35?

Q5. If p is a factor of x, and x is a factor of y, is p a factor of y?

Q6. If x and y are multiples of 12, is $(3x + 7y)$ a multiple of 12?

Q7. If $\frac{x}{y}$ an odd integer, is x odd? Given that x and y are positive integers.

Q8. If 4,611 is divisible by 159, is it also divisible by 53?

Q9. If 9 is a factor of both x and y, is $(x + y)$ divisible by 9?

4.4 Simplification

Q1. Simplify $4 - [6 - 3^4(6 \div 3 + 1)]$.

Q2. $8 \div \left(\frac{8}{7} \times 25\% \text{ of } 28\right)$.

Q3. Simplify $(9.75 \times 10^{-2}) - (25 \times 10^{-3})$.

Q4. What is the value of p, if $3\sqrt[3]{p} + 92 = -100$.

Q5. Simplify $(7.75 \times 10^{2}) \div (25 \times 10^{-3})$.

Q6. Simplify $\sqrt[9]{0.000000512}$.

Q7. If $m \neq 4$ and $m \neq 3$, what is the value of $\left[\frac{4m^2 - 64}{m - 4} - \frac{4m^2 - 36}{m - 3}\right]$?

Q8. If $x = 0.999$, $y = \sqrt{0.999}$, and $z = (0.999)^2$, which among x, y and z is the largest and the smallest?

Q9. Between which two integers, the value of $\sqrt{30}$ lies?

Q10. Between which two integers the value of $\sqrt[3]{50}$ lies?

4.5 Exponents

Q1. Simplify $\frac{3^{t+3} + 3^t}{2 \times 3^t}$.

Q2. Simplify $64^{\frac{1}{3}} + 16^{\frac{1}{2}} + \frac{4}{(16)^0}$.

Q3. Simplify $\frac{(-4)^3 \times (2^5)^3}{-4^{-3}}$.

Q4. The expression $\left(\frac{-9}{4}\right)^{-2}$ is equivalent to the following EXCEPT.

A. $\left(\frac{-4}{9}\right)^2$

B. $\left(\frac{4}{9}\right)^2$

C. $\left(\frac{3}{2}\right)^{-4}$

D. $\left(\frac{2}{3}\right)^4$

E. $\left(\frac{2}{3}\right)^{-4}$

Q5. If $4^{x-1} \times (0.5)^{3-2x} = \left(\frac{1}{8}\right)^{-x}$, what is the value of x?

Q6. If $3^{m+n} = \sqrt{p - 4}$, and $m = 2 - n$, what is the value of p?

Q7. If n is an integer, and 0.00876×10^n is greater than 1000, what is the least possible value of n?

Q8. If a number $46,567 \div 10^{-n}$ lies between 4 and 400, what is set of integer values of n?

Q9. If a number -0.046567×10^n lie between 3 and -500, what is set of integer values of n?

4.6 Factors/Multiples and LCM/HCF

Q1. If $11x$ is a multiple of 16,500, is x a multiple of 70?

Q2. If 21 and 10 are factors of a number n, is 14 also a factor of n?

Q3. X and Y are both two-digit numbers such that $X > Y$. If X and Y contain the same digits, but in reverse order, which of the following must be a factor of $(X - Y)$?

A. 4

B. 5

C. 6

D. 9

E. 10

F. 11

Q4. If $x = 35n$, and $y = 55n$, whether n is prime, what is the greatest common factor of x and y?

4.7 Fraction

Q1. If x equals $\frac{45}{47}$ of $\frac{47}{43}$, and y equals the reciprocal of $\frac{45}{43}$, then what is the value of $\frac{x}{y}$?

Q2. Brian ate 1/3, and Suzy ate some fraction of the cookies, and then Rachel ate 1/2 of the remaining cookies. If 1/6 of the cookies are left in the jar, what fraction of the cookies was eaten by Suzy?

Q3. For each of the following choices, which of the two fractions is larger?

A. $\frac{45}{44}, \frac{91}{92}$

B. $\frac{45}{44}, \frac{93}{92}$

C. $\frac{44}{45}, \frac{91}{92}$

D. $\frac{83}{81}, \frac{114}{111}$

4.8 Rationalization of factions

Q1. Rationalize $\frac{4}{3\sqrt{2}} - \frac{2}{\sqrt{2}}$.

Q2. Simplify $\frac{\sqrt{50}}{\sqrt{10}} \times \frac{\sqrt{18}}{\sqrt{12}}$.

4.9 Linear and Quadratic Equations

Q1. For the following two linear equations, find out the values of x and y.

$$x - 2y - 1 = 0 \;\&$$
$$-3y - x - 1 = 0$$

Q2. For the following two linear equations, find out the values of x and y.

$$\frac{x}{2} - \frac{y}{3} + 4 = 0 \;\&$$
$$x - \frac{y}{2} = -7$$

Q3. For $2x^2 + 7x + 3 = 0$, what are the values of x?

Q4. For $x - 3\sqrt{x} + 2 = 0$, what are the values of x?

Q5. For $x^{2/3} + x^{1/3} - 2 = 0$, what are the values of x?

Q6. For what values of m does the quadratic equation $x^2 + 2(m - 4)x + 2m = 0$ have equal roots.

Q7. If $(4x - 1)^2 = 81$, what is x?

Q8. If $(a - 3)(a + 2) = (a + 3)(a - 2)$, how many unique values a can have?

Q9. If $a^3 - a^2 - a + 1 = 0$, how many unique values a can have?

4.10 Absolute numbers

Q1. If $|10x| - 20 = 0$, what is x?

Q2. What is the value of $\left[\frac{-|-3| \times -|3|}{(-|-3|)^2}\right]$?

Q3. If $|2x - 5| = 8$, what is x?

4.11 Inequalities

Q1. How many values x might have for the inequality $1 \leq x < 3$?

Q2. If $\frac{1}{p} > 1$, what is the range of value of p?

Q3. What do the following inequalities suggest about the values of a and b?

A. $ab > 0$

B. $ab < 0$

C. $ab = 0$

D. $\frac{a}{b} > 0$

E. $\frac{a}{b} < 0$

Q4. For which values of x, will the inequality $|2x - 1| < 5$ be true?

Q5. Solve the $|x + 2| < 7$ inequality for x.

Q6. Solve the inequality $|1 - 2x| < 9$ for x.

Q7. Solve the inequality $\frac{5x - 1}{-6} \leq \frac{1 - 2x}{3}$ for x.

Q8. If $1 < x < 3$, then which of the following could be true?

I. $x^2 < 2x$

II. $x^2 = 2x$

III. $x^2 > 2x$

A. I only

B. II only

C. III only

D. I and II only

E. All three

Q9. What does the inequality $ab^2c < 0$ suggest for the values of a, b, and c?

Q10. If $8 > x > 3$ and $4 < y < 6$, what is the range of values of $(x - y)$?

Q11. For what values of x, the inequality $x^2 > -6x - 5$ exists?

Q12. For what values of x, the inequality $x^2 > 6x - 5$ exists?

4.12 Percents

Q1. 24 is 8% of what number?

Q2. If an industrial worker rejects 0.03% of screws as defective, how many screws does he examine to reject 300 screws?

Q3. What percent of 17 is 30 percent of 289?

Q4. How much you have to pay for an item priced \$800, and discounted at 40%?

Q5. An item was sold at \$100; it was bought for \$120. What is the percent loss?

Q6. An item was bought at \$100 after a discount of \$20 on Marked Price. What is the percent discount on the Marked Price?

Q7. If X is more than Y by 110%, and Z is less than X by 10%, then X is what % of Z?

Q8. If a number is first increased by 20%, and then decreased by 20%, is the resulting number equal to the original number?

Q9. A shopkeeper increased the price of his articles by 25%. What percent discount should he offer so that he does not run into losses?

4.13 Ratio & Proportion

Q1. If 0.25 : 5 :: x : 4, then x = ?

Q2. Two numbers are respectively 20% and 50% greater than the third number. What is the ratio of the two numbers?

Q3. If \$529 be divided into three parts, proportional to $\frac{1}{2} : \frac{2}{3} : \frac{3}{4}$, then what is the value of the third part?

Q4. A and B together have \$240. If $\frac{3}{5}$ of A's amount is equal to $\frac{2}{5}$ of B's amount, how much does B have?

Q5. Two number are in the ratio of 3 : 5. If 9 is subtracted from each, new numbers are in the ratio of 12 : 23. What is the value of the larger number?

Q6. If $A : B : C = 2 : 3 : 4$, then find $\frac{A}{B} : \frac{B}{C} : \frac{C}{A}$.

Q7. A 60 liters mixture has milk and water in the ratio of 2 : 1. If the ratio is to be reversed, then what quantity of water should be added?

Q8. The sum of three numbers is 190. If the ratio of the first to second is 2 : 3 and that of the second to the third is also 2 : 3, then the third number is

Q9. The salaries of A, B, and C are of in the ratio of 2 : 3 : 5. If the increments of 15%, 10%, and 20% are done to their respective salaries, find the new ratio of their salaries.

Q10. If three numbers are in the ratio of 2 : 3 : 5, and the sum of their squares is 1,368, what is the value of the first number?

4.14 Simple Interest

Q1. Find the simple interest on \$8,000 borrowed at $\frac{50}{3}$% for 27 months.

Q2. Brian took a loan for 6 years at the rate of 5% per annum on simple interest, If the total interest paid was \$1,230, what was the principal?

Q3. If a sum becomes four times of itself in 15 years, find simple rate of interest.

Q4. Brian borrows \$5,000 for 2 years at 4% p.a. simple interest. He immediately lends whole amount to Suzy at $\frac{25}{4}$% p.a. for 2 years. Find the gain for Brian.

Q5. A lent \$5,000 to B for 2 years and \$3,000 to C for 4 years on simple interest at the same rate of interest and received \$2,200 in all from both of them as interest. What is the rate of interest per annum?

4.15 Compound Interest

Q1. Find the compound interest on \$7,500 at 4% per annum for 2 years, compounded annually.

Q2. Simple interest on a certain sum for 3 years at 8% per annum is equal to half the compound interest on \$4,000 for 2 years at 10% per annum. What is the sum placed on simple interest?
Q3. If the difference between simple interest and compound interest compounded annually on a certain sum of money for 2 years at 4% per annum is \$1, find the sum.

Q4. What will be the equivalent effective annual rate of interest corresponding to nominal rate of 20% per annum compounded half yearly?

Q5. If a sum of money invested at compound interest amounts to \$800 in 3 years and to \$840 in 4 years, what is the rate of interest per annum?

4.16 Functions

Q1. Given $f(x) = 2x - 7$, for what value of x, does $3f(x) - 3 = f(2x - 6)$?

Q2. For all positive integers $p, [p] = -|\sqrt{p}|$, when p is odd and $[p] = \frac{p}{5}$, when p is even. What is the value of $[25] \times [10]$?

Q3. If $[m]$ is the greatest integer less than or equal to m, what is the value of $[-2.2] + [2.6] + [6.2]$?

Q4. An operation # is defined by the equation $a\#b = \frac{(a+b)}{(a^2+b^2)}$, for all numbers a and b such that $a \neq 0$. If $a\#c = 0$, what is the value of c in terms of a?

Q5. Given $f(x) = \frac{2x}{2x-1}$, for what value of x, does $f(f(x)) = -1$?

Q6. Given $f(x) = \frac{2x}{2x-1}$ and $g(x) = \frac{2x-1}{2x}$, for what value of x, does $f(g(x)) = 3$?

Q7. If $f(x) = 2x + \sqrt{x}$, what is the value of $f(p^2 - 10p + 25)$?

Chapter 5

Answer Key

Numbers and Digits

1.1: Irrational; 1.2: Irrational; 1.3: Rational; 1.4: Rational;

1.5: Irrational

2: $\frac{646}{99}$ 3: $\frac{524}{999}$ 4: 0 5: $n = \{0, \ldots, 5\}$

6: 78

Even/Odd and Consecutive Integers

1: Indeterminable 2: Indeterminable 3: Yes 4: No

5: Yes 6: Indeterminable 7: Indeterminable 8: A, B, D, & E

9: Indeterminable 10: Yes

Divisibility

1: 37 2: Yes 3: Indeterminable 4: Yes

5: Yes 6: Yes 7: Indeterminable 8: Yes

9: Yes

Simplification

1: 241
2: 1
3: 0.0725
4: -2^{18}
5: 31,000
6: 0.2
7: 4
8: y: Largest
z : Smallest
9: 5 & 6
10: 3 & 4

Exponents

1: 14
2: 12
3: 2^{27}
4: E
5: 5
6: 85
7: 6
8: $\{-3, -4\}$
9: $\{2, 3, 4\}$

Factors/Multiples and LCM/HCF

1: Cannot say
2: Yes
3: D
4: $5n$

Fraction

1: $\left(\frac{45}{43}\right)^2$
2: 1/3
3.A: $\frac{45}{44}$
3.B: $\frac{45}{44}$
3.C: $\frac{91}{92}$
3.D: $\frac{114}{111}$

Rationalization of fraction

1: $-\frac{\sqrt{2}}{3}$
2: $\frac{\sqrt{30}}{2}$

Linear and Quadratic Equations

1: $x = \frac{1}{5}$ & $y = -\frac{2}{5}$

2: $x = -4$ & $y = 6$

3: -3 or $-\frac{1}{2}$

4: 4 or 1

5: -8 or 1

6: 8 or 2

7: $\frac{5}{2}$ or -2

8: One

9: Two

Absolute numbers

1: $x = \pm 2$

2: 1

3: $-\frac{3}{2}$ or $\frac{13}{2}$

Inequalities

1: Infinite

2: $0 < p < 1$

3.A: $a > 0$ & $b > 0$ or $a < 0$ & $b < 0$

3.B: $a > 0$ & $b < 0$ or $a < 0$ & $b > 0$

3.C: $a = 0$ &/or $b = 0$

3.D: $a > 0$ & $b > 0$ or $a < 0$ & $b < 0$

3.E: $a > 0$ & $b < 0$ or $a < 0$ & $b > 0$

4. $-2 < x < 3$

5. $-9 < x < 5$

6. $-4 < x < 5$

7. $x \geq -1$

8. E

9. $a > 0$ & $c < 0$ or $a < 0$ & $c > 0$

10. $4 > x - y > -3$

11. $x > -1$ or $x < -5$

12. $x < 1$ or $x < 5$

Percents

1: 300	2: 1,000,000	3: 510	4: \$480	5: 16.67%
6: 16.67%	7: 111.11%	8: No; less by 4%	9: 20%	

Ratio & Proportion

1: 1/5	2: 4:5	3: 207	4: 144	5: 55
6: 8:9:24	7: 60 liters	8: 90	9: 23 : 33 : 60	10: 12

Simple Interest

1: \$3,000	2: \$4,100	3: 20%	4: \$225	5: 10%

Compound Interest

1: \$612	2: \$1750	3: \$625	4: 21% p.a.	5: 5%

Functions

1: $x = 5/2$	2: -10	3: 5	4: $-a$	5: $-1/6$
6: 2	7: $2p^2 - 19p + 45$; $2p^2 - 21p + 55$.			

Chapter 6

Solutions

6.1 Numbers and Digits

Q1.

(1) Irrational; π is an irrational number as the decimal does not terminate and has no repeated pattern.

(2) Irrational; decimal does not terminate and has no repeated pattern.

(3) Rational; decimal has a repeated pattern.

(4) Rational; decimal has a repeated pattern.

(5) Irrational; exponent of an irrational number would be an irrational number.

Q2.

- Step 1: Say $x = 6.\overline{52}$.
- Step 2: Now multiply x with a minimum value of exponent of 10 such that recurring part of the number x becomes an integer. So, $100x = 652.\overline{52}$.
- Step 3: Now deduct $x = 6.\overline{52}$ from $100x = 652.\overline{52}$; we get $99x = 646$ or $x = \frac{646}{99}$; a rational fraction.

Q3.

- Step 1: Say $x = 0.\overline{524}$.
- Step 2: Now multiply x with a minimum value of exponent of 10 such that recurring part of the number x becomes an integer. So, $1000x = 524.\overline{524}$.
- Step 3: Now deduct $x = 0.\overline{524}$ from $1000x = 524.\overline{524}$; we get $999x = 524$ or $x = \frac{524}{999}$; a rational fraction.

Q4. 0.

The numbers a, b, c, and d can have any number of digits; however for this question, it is not of any importance to know how many digits each number has as while adding numbers, we add unit digits first; so the addition of unit digits of a, b, c, and d would be $2 + 2 + (-2) + (-2) = 4 - 4 = 0$.

Alternatively, you can assume some values for a, b, c, and d. Say $a = 2,452, b = 32, c = -82$, and $d = -2$. It is clear that the units digits of the sum of a, b, c, and d would be 0.

Q5.

It is to be noted that n must be less than '5' as if it were '5' or more than '5', the tenth digit '4' would be rounded off to '5' and the number would be '25.5'. Remember that if the digit to the left of '5' is even (In this case it is '4'), it is not increased by '1'. So, to keep the number less than '25.5', the set of $n = \{0, 1, 2, 3, 4\}$.

Q6.

$77,567,893 \times 10^{-6} = 77.567893$. Since we have to round off to the nearest whole number, the number 77.567893 would be '78' since the digit on its tenth place is '5,' which means that we must increase the digit to its left by '1' or the digit on tens place '7' becomes '8'.

6.2 Even/Odd and Consecutive Integers

Q1.

Sum of two integers can be even in two situations:

(1) Both a and b are even. For example, $a = 2$, and $b = 4$. Sum = 2 + 4 = 6 (Even). This gives $ab = 8$; an even number.

(2) Both a and b are odd. For example, $a = 1$, and $b = 3$. Sum = 1 + 3 = 4 (Even). This gives $ab = 3$; an odd number. So it is indeterminable whether ab is odd or even.

Q2.

So, the situation is: Is (p + even) even? It will depend on the nature of p; if it is even, (p + even) would be even, else odd.

Given is pq = even; but this will not help us anyway as for pq to be even p can be even and q can be odd, or vice-versa. So the nature of $(p + r)$ is indeterminable.

Q3.

Consecutive integers may start from an even integer, say {2, 3, 4}, or an odd integer, say {3, 4, 5}. In either case, at least one of the consecutive integers, x, y, and z is even. And we know that the product of an even integer either with an even or an odd integer is an even integer; therefore, the product xyz must be even.

Q4. No.

So the situation is $z(x + y)$ = Odd.(Even + Odd) = Odd.Odd = Odd. The answer is No. $z(x + y)$ is odd.

Q5. Yes.

$z(x + y)$ is even. The situation is $z(x + y)$ = Even.$(x + y)$ = Even.(Odd) or Even.(Even); however both are even; it is insignificant to know what the nature of $(x + y)$ is.

Q6. Indeterminable.

The situation is $z(x + y)$ = Odd.(x+Odd). Since Odd.Even = Even and Odd.Odd = Odd, it is necessary to know the nature of (x+Odd); if x is even, (x+Odd) is odd, and if x is odd, (x+Odd) is even. So it is indeterminable.

Q7. Indeterminable.

Follow the same reasoning as used in the above questions.

Q8.

Let us abbreviate Even as 'E' and Odd as 'O'.

We know that Even × Even = Even & Even × Odd = Even, and Even + Even = Even & Even + Odd = Odd.

A. $n^2 + 3$: The expression is equivalent to $E^2 + O = E.E + O = E + O = O$; it is an **odd number**.

B. $3n^2 - 3$: The expression is equivalent to $O.E^2 - O = O.E - O = E - O = O$; it is an **odd number**.

C. $n^3 + 2$: The expression is equivalent to $E^3 + E = E + E = E$; it is an even number.

D. $2n + 3$: The expression is equivalent to $E.E + O = E + O = O$; it is an **odd number**.

E. $-3n + 3$: The expression is equivalent to $-O.E + O = -E + O = O$; it is an **odd number**.

F. $n(n^4 - 3n^3 + 14n^2 - 4)$: The expression is equivalent to E. (Unknown number: Even or Odd) $= E.(O \text{ or } E) = E$; multiplication of any number with an even number is always even.

G. $\sqrt{n^2 + 12}$: The expression is equivalent to $\sqrt{E^2 + E} = \sqrt{E + E} = \sqrt{E}$; square root of an even number may be even or real, but cannot be an odd number.

The correct answers are options A, B, D, & E.

Q9.

There would be two situations:

(1) The five consecutive integers starting with an even integer are {2, 3, 4, 5, 6}; the sum is even.

(2) The five consecutive integers starting with an odd integer are {1, 2, 3, 4, 5}; the sum is odd. So it is indeterminable.

Q10.

Again there would be two situations:

(1) The four consecutive integers starting with an even integer are {2, 3, 4, 5}; the sum is even.

(2) The four consecutive integers starting with an odd integer are {1, 2, 3, 4}; the sum is even.

So the sum is even.

Remember that sum of even number of consecutive integers is always even.

6.3 Divisibility

Q1.

Say the number is $(10,000n + 37)$; where n is a positive integer. Since $10,000n$ is divisible by 1,000, the question comes down to whether 37 is divisible by 1000. Since it is not divisible by $1,000$, the remainder would be 37 only.

Q2.

Since a is divisible by 7, a would be $7n$, where n is a positive integer; similarly, since a is divisible by 18, a would be $18m$; where m is a positive integer. But a can be any one value between $7n$ or $18m$. Since 7 and 18 are co-prime to each other, a should be $7 \times 18p$; where p is a positive integer.

Two numbers are co-prime to each other if no factor other than '1' is common between them.

Now, is $\frac{a}{42}$ an integer?

Plugging in the value of a, we get $\frac{a}{42} = \frac{7 \times 18p}{2.3.7} = \frac{\cancel{7} \times \cancel{2} \times \cancel{3} \times 3p}{\cancel{2}.\cancel{3}.\cancel{7}} = 3p$ (an integer); so $\frac{a}{42}$ is an integer.

Q3.

For 20 to be a factor of x given that 70 is a factor of x, 20 must be a factor of 70—which is not true. So whether 20 is a factor of x is indeterminable. Note that 20 may or may not be a factor of x. Let us see how it may be a factor of x.

Say $x = 70n$; where n is a positive integer. For 20 to be a factor of x, $\frac{70n}{20}$ must be an integer. After reducing $\frac{70n}{20}$, we get $\frac{7n}{2}$. If n is a multiple of 2, 20 is a factor of x, else not.

Another approach would be: say if $x = 140$ (a multiple of 70), then 20 is a factor of x; however if $x = 210$ (a multiple of 70), then 20 is NOT a factor of x. So it is inconclusive!

Q4.

The answer is Yes. Since prime factors of 35 = {5, 7} are common either with the prime factors of 15 = {3, 5} or with the prime factors of 14 = {2, 7}, p is divisible by 35.

Another approach would be: Since p is a multiple of 15 and 14, p must be a multiple of all the factors of 15 and 14. This means that $p = 2.3.5.7.n = 35.(6n)$; where n is a positive integer. It is clear that p is a multiple of 35 or is divisible by 35.

Q5.

Say $x = pn$ and $y = xm$; where n & m are positive integers. So $y = pnm$, and we can deduce that p a factor of y.

Q6.

Say $3x + 7y = 3(12n) + 7(12m) = 12(3n + 7m)$; where n & m are positive integers. It is clear that $12(3n + 7m)$ is a multiple of 12.

Q7.

$\frac{x}{y}$ can be odd in two situations:

(1) Both x and y are even, and x is a multiple of an odd number. For example, $x = 12$, and $y = 4$, so $\frac{x}{y} = \frac{12}{4} = 3$ (an odd number).

(2) Both x and y are odd. For example, $x = 21$, and $y = 7$, so $\frac{x}{y} = \frac{21}{7} = 3$ (an odd number). So it is indeterminable whether x is odd or even.

Q8. Yes.

Since 53 is factor of a 159, ($159 = 53\times3$), 4611 will also be divisible by 53. Remember that if a divisor x divides a number y completely, all the factors of x will also divide y completely.

Q9.

Since 9 is a factor of x, as well as y, $x = 9n$ and $y = 9m$. So $x + y = 9n + 9m = 9(n + m)$, which is divisible by 9 or $(x + y)$ is divisible by 9.

6.4 Simplification

Q1.

$4 - [6 - 3^4(6 \div 3 + 1)]$

=> $4 - [6 - 3^4(\frac{6}{3} + 1)]$; innermost parenthesis first, and 'division' before 'addition'

=> $4 - [6 - 3^4(2 + 1)]$; performing division

=> $4 - [6 - 3^4(3)]$; performing addition

=> $4 - [6 - 81.(3)]$; expanding exponent

=> $4 - [6 - 243]$; performing multiplication

=> $4 - [-237]$; performing subtraction

=> $4 + 237$; performing multiplication of two negatives

=> 241; performing addition

Q2.

$8 \div \left(\frac{8}{7} \times 25\% \text{ of } 28\right)$

=> $8 \div \left(\frac{8}{7} \times \frac{25}{100} \times 28\right)$

=> $8 \div \left(\frac{8}{7} \times \frac{\cancel{25}}{\cancel{100}^{4}} \times \cancel{28}^{7}\right)$

=> $8 \div \left(\frac{8}{\cancel{7}} \times \cancel{7}\right)$

= > $8 \div 8$

=> 1.

Q3.

$(9.75 \times 10^{-2}) - (25 \times 10^{-3})$

=> $0.0975 - 0.025$

$\Rightarrow 0.0725$

Alternatively, we can simplify in the following way.

$(9.75 \times 10^{-2}) - (25 \times 10^{-3})$

$\Rightarrow (9.75 \times 10^{-2}) - \left(\frac{25}{10} \times 10^{-2}\right)$

$\Rightarrow 10^{-2}[9.75 - 2.5]$

$\Rightarrow 10^{-2}[7.25]$

$\Rightarrow \frac{7.25}{100}$

$\Rightarrow 0.0725$

Q4.

$3\sqrt[3]{p} + 92 = -100$

$\Rightarrow 3\sqrt[3]{p} = -192$

$\Rightarrow \sqrt[3]{p} = -\frac{192}{3}$

$\Rightarrow \sqrt[3]{p} = -64$

$\Rightarrow p = (-64)^3 = -2^{18}$

Q5.

$(7.75 \times 10^2) \div (25 \times 10^{-3})$

$\Rightarrow \frac{7.75 \times 10^2}{25 \times 10^{-3}}$

$\Rightarrow \frac{775}{25 \times 10^{-3}}$

$\Rightarrow \frac{\cancel{775}^{31} \times 10^3}{\cancel{25}}$

$\Rightarrow 31,000.$

Q6.

$\sqrt[9]{0.000000512}$

=> $(0.000000512)^{1/9}$

=> $(512 \times 10^{-9})^{1/9}$

=> $(512)^{1/9} \times 10^{(-9\times1/9)}$

=> $2^{(9\times1/9)} \times 10^{-1}$

=> 2×10^{-1}

=> 0.2.

Q7.

$\frac{4m^2 - 64}{m - 4} - \frac{4m^2 - 36}{m - 3}$ can be simplified as

$= \frac{4(m^2 - 16)}{m - 4} - \frac{4(m^2 - 9)}{m - 3}$

$= \frac{4(m + 4)\cancel{(m - 4)}}{\cancel{(m - 4)}} - \frac{4(m + 3)\cancel{(m - 3)}}{\cancel{(m - 3)}}$

$= 4m + 16 - 4m - 12$

$= 4$

Q8.

Note that the base number $0.999 < 1$. Remember that...

(1) n^{th} root of a number less than 1 is always greater than the number itself; so $0.999 < \sqrt{0.999}$.

(2) Squares, cubes, and other high value powers of a number less than 1 are always smaller than the number itself; so $0.999 > (0.999)^2$.

It means that $z < x < y$. So, y is the largest, and z the smallest.

Apposite is true for a number greater than 1.

Q9.

First, find out which two perfect square numbers are closest to '30'; these are 25 and 36. So we can say that $(25 = 5^2) < 30 < (36 = 6^2)$ or $5 < \sqrt{30} < 6$. So $\sqrt{30}$ lies between 5 and 6.

Q10.

First, find out which two perfect cube numbers are closest to '50'; these are 27 and 64. So we can say that $(27 = 3^3) < 50 < (64 = 4^3)$ or $3 < \sqrt[3]{50} < 4$. So $\sqrt[3]{50}$ lies between 3 and 4.

6.5 Exponents

Q1.

$$\frac{3^{t+3} + 3^t}{2 \times 3^t}$$

$$= \frac{3^t.3^3 + 3^t}{2 \times 3^t}$$

$$= \frac{3^t.(3^3 + 1)}{2 \times 3^t}\text{; taking } 3^t \text{ as common.}$$

$$= \frac{\cancel{3^t}.(27 + 1)}{2 \times \cancel{3^t}}$$

$= 14.$

Q2.

$$64^{\frac{1}{3}} + 16^{\frac{1}{2}} + \frac{4}{16^0}$$

$$= (2^6)^{\frac{1}{3}} + (2^4)^{\frac{1}{2}} + \frac{4}{1}\text{; since } a^0 = 1$$

$$= 2^{6\times1/3} + 2^{4\times1/2} + 4$$

$$= 2^2 + 2^2 + 4$$

$$= 4 + 4 + 4 = 12.$$

Q3.

$\dfrac{(-4)^3 \times (2^5)^3}{-4^{-3}}$

$= \dfrac{(-1)^3.4^3.2^{5\times3}}{-\dfrac{1}{4^3}}$

$= \dfrac{\cancel{-1}.4^3.4^3.2^{15}}{\cancel{-1}}$; since $(-1)^{odd} = -1$, and $(-1)^{even} = +1$

$= 4^6.2^{15}$

$= (2^2)^6.2^{15}$

$= 2^{2\times6}.2^{15}$

$= 2^{12}.2^{15}$

$= 2^{12+15}$

$= 2^{27}$

Q4.

Let us solve the expression and get the desired equivalents.

$= \left(\dfrac{-9}{4}\right)^{-2} = \left(\dfrac{-4}{9}\right)^{2}$; when the exponent changes sign, numerator and denominator swap.

Option A is an equivalent expression.

$= \left(\dfrac{-4}{9}\right)^{2} = \left(\dfrac{4}{9}\right)^{2}$; 'negative number' raised to an even exponent is always positive.

Option B is an equivalent expression.

$=> \left(\dfrac{-9}{4}\right)^{-2} = \left(\dfrac{9}{4}\right)^{-2}$; 'negative number' raised to an even exponent is always positive.

$= \left(\dfrac{3^2}{2^2}\right)^{-2} = \left(\dfrac{3}{2}\right)^{-2\times2} = \left(\dfrac{3}{2}\right)^{-4}$.

Option C is an equivalent expression.

Similarly, $\left(\dfrac{3}{2}\right)^{-4} = \left(\dfrac{2}{3}\right)^{4}$; when the exponent changes sign, numerator and denominator swap.

Option D is an equivalent expression.

So the correct answer is option E.

Q5.

$4^{x-1}.(0.5)^{3-2x} = \left(\frac{1}{8}\right)^{-x}$

$=> (2^2)^{x-1}.\left(\frac{5}{10}\right)^{3-2x} = \left(\frac{1}{2^3}\right)^{-x}$

$=> 2^{2x-2}.\left(\frac{\cancel{5}}{\cancel{10}^{\,2}}\right)^{3-2x} = (2^{-3})^{-x}$

$=> 2^{2x-2}.\left(\frac{1}{2}\right)^{3-2x} = 2^{-3x-x}$

$=> 2^{2x-2}.(2^{-1})^{3-2x} = 2^{-3x-x}$

$=> 2^{2x-2}.2^{-3+2x} = 2^{3x}$

$=> 2^{2x-2-3+2x} = 2^{3x}$

$=> 2^{4x-5} = 2^{3x}$

It means that $4x - 5 = 3x$ => $\boldsymbol{x = 5}$; if bases are equal, exponents would be equal.

Q6.

Plug in the value of $m = 2 - n$ in 3^{m+n}, we get,

$3^{2-n+n} = \sqrt{p-4}$

$=> 3^2 = \sqrt{p-4}$

$=> 9 = \sqrt{p-4}$

$=> 9^2 = (\sqrt{p-4})^2$; squaring both the sides

$=> 81 = p - 4$

$=> p = 85$

Q7.

For the number 0.00876 to be greater than 1000, decimal must be moved six places to the right and the resulting number would be $8760 > 1000$. This means that the number must be multiplied with 10^6. Hence the least possible value of $n = 6$.

Q8.

Only the numbers 4.6567 and 46.567 would lie between 4 and 400. The number $46567 \div 10^{-n}$ can be written as 46567.0×10^n. To get the number 4.6567, we must move the decimal four places to its left in 46567.0, which means that the number must be multiplied with 10^{-4} or $n = -4$.

Similarly, to get the number 46.567, we must move the decimal three places to its left in 46567.0, which means that the number must be multiplied with 10^{-3} or $n = -3$. Hence the set of $n = \{-3, -4\}$.

Q9.

Numbers -4.6567, -46.567, and -465.67 would lie between -3 and -500. Which means that we must move the decimal two, three, and four places, respectively to its right to get these numbers. Hence the set of $n = \{2, 3, 4\}$.

6.6 Factors/Multiples and LCM/HCF

Q1.

We cannot conclude. We can deduce that $11x = 16500n$; where n is a positive integer. So $\cancel{11}x = \cancel{16500}^{1500} n \Rightarrow x = 1500n$. It is clear that 1500 does not have any factor '7'. If $n = 7$, the answer is yes, else not! Thus, the answer is indeterminable.

Q2.

Since both the prime factors of 14 (= 2 & 7) are common with either the prime factors of 21 (= 3 & 7) or the prime factors of 10 (= 2 & 5), 14 is also a factor of n.

Alternatively, since 21 (= 3 & 7), and 10 (= 2 & 7) are factors of n, so $n = 2.3.7.m = 210m$; where m is a positive integer. It is clear that '14' is factor of '210', or 14 also a factor of n.

Q3.

A two-digit number can be written as $10x + y$; where x is the tens digits and y is units digit. So $X = 10x + y$ and $Y = 10y + x$, as the digits are reversed.

=> $X - Y = (10x + y) - (10y + x) = 9x - 9y = 9(x - y)$; it is clear that 9 is factor of $(X - Y)$. So the correct option is D.

Q4.

Greatest common factor among numbers is also called HCF or GCD. The question asks us to find out the HCF of $35n$ and $55n$. Since $35n = 5.7.n$ and $55n = 5.11.n$ have '5' and 'n' common, the HCF would be $5n$ or the greatest common factor of x and y is '$5n$'.

Alternatively,

$x = \not{5}.7.\not{n};$
$y = \not{5}.11.\not{n};$

HCF = $5.n$

The answer is $5n$.

6.7 Fraction

Q1.

$$x = \frac{45}{47} \times \frac{47}{43} = \frac{45}{43} \ \&$$

$$y = \text{reciprocal of } \frac{45}{43} = \frac{43}{45}$$

$$\text{So, } \frac{x}{y} = \frac{\left(\frac{45}{43}\right)}{\left(\frac{43}{45}\right)} = \frac{45}{43} \times \frac{45}{43} = \frac{45^2}{43^2} = \left(\frac{45}{43}\right)^2$$

$$\Rightarrow \frac{x}{y} = \left(\frac{45}{43}\right)^2$$

Q2.

Say Suzy ate $\left(\frac{1}{n}\right)^{\text{th}}$ fraction of cookies, then the fraction of cookies left after eaten up by Brian and Suzy $= 1 - \frac{1}{3} - \frac{1}{n} = \frac{2n - 3}{3n}$.

Since 1/2 of $\frac{2n-3}{3n}$ was eaten by Rachel, Fraction of cookies left in the jar = $\frac{1}{2}.\left(\frac{2n-3}{3n}\right)$.

Or, $\frac{1}{2}.\left(\frac{2n-3}{3n}\right) = \frac{2n-3}{6n}$

=> $\frac{2n-3}{6n} = \frac{1}{6}$; given that 1/6 of the cookies are left in the jar

=> $n = 3$. Suzy ate $1/n = 1/3$ fraction of cookies.

Let us cross check...

Fraction of cookies left after eaten up by Brian and Suzy = $1 - \frac{1}{3} - \frac{1}{3} = \frac{1}{3}$.

It means that Rachel ate $\frac{1}{2}$ of $\frac{1}{3} = \frac{1}{6}$ fraction of cookies.

Fraction of cookies left in the jar = $1 - \frac{1}{3} - \frac{1}{3} - \frac{1}{6} = \frac{1}{6}$, which is equal to the fraction given in the equation. So our calculation was correct.

Alternatively...

Say total number of cookies were 6. We chose a smart number '6' as from the question, we know that we have to deal with the fractions: 1/3 and 1/2, so the LCM is 6, a convenient number.

Total number of cookies = 6;

Total number of cookies eaten by Brian = 1/3 of 6 = 2;

Say, the total number of cookies eaten by Suzy = x;

So, the remaining number of cookies after eaten by Brian and Suzy = $6 - 2 - x = 4 - x$;

So, the total number of cookies eaten by Rachel = 1/2 of $4 - x = \frac{4-x}{2}$;

We know that the total number of cookies left in the jar = 1/6 of 6 = 1.

So, $x + \frac{4-x}{2} = 2 + 1 = 3 \Rightarrow x = 2$; or Suzy ate $x/6 = 2/6 = 1/3$ fraction of cookies.

Q3.

A. $\frac{45}{44}, \frac{91}{92}$

The numerator of fraction $\frac{45}{44}$ is larger than its denominator, hence $\frac{45}{44} > 1$; whereas the numerator of fraction $\frac{91}{92}$ is smaller than its denominator, hence $\frac{91}{92} < 1$. So, $\frac{45}{44} > \frac{91}{92}$.

B. $\frac{45}{44}, \frac{93}{92}$

$\frac{45}{44}$ can be written as $\frac{45+1}{44} = 1\frac{1}{44}$; similarly $\frac{93}{92} = 1\frac{1}{92}$. The integer part of both the mixed fractions are equal (1) and the fraction part of $1\left|\mathbf{\frac{1}{44}}\right|$ is greater than the fraction part of $1\left|\mathbf{\frac{1}{92}}\right|$; $\left(\mathbf{\frac{1}{44}} > \mathbf{\frac{1}{92}}\right)$; smaller the denominator, larger is the fraction. So $\frac{45}{44} > \frac{93}{92}$.

C. $\frac{44}{45}, \frac{91}{92}$

$\frac{44}{45}$ can be written as $\frac{44-1}{44} = 1 - \frac{1}{44}$; similarly $\frac{91}{92} = 1 - \frac{1}{92}$. The integer part of both the mixed fractions are equal (1) and though the absolute vale of the fraction part of $1\left|-\mathbf{\frac{1}{44}}\right|$ is greater than the fraction part of $1\left|-\mathbf{\frac{1}{92}}\right|$, it is smaller since it is negative. $\left(-\mathbf{\frac{1}{44}} < -\mathbf{\frac{1}{92}}\right)$; So $\frac{44}{45} < \frac{91}{92}$.

D. $\frac{83}{81}, \frac{114}{111}$

$\frac{83}{81}$ can be written as $\frac{81+2}{81} = 1\frac{2}{81}$; similarly $\frac{114}{111} = 1\frac{3}{11}$. The integer part of both the mixed fractions are equal (1), but fraction parts cannot be compared as neither the numerator nor the denominator of the first fraction is equal to the relevant counterparts of the second fraction.

So the question reduces to: Which of two fractions $\frac{2}{81}$ or $\frac{3}{111} is larger$?

The fraction $\frac{2}{81}$ can be written as $\frac{1}{81/2} = \frac{1}{\# > 40}$; similarly, the fraction $\frac{3}{111}$ can be written as $\frac{1}{111/3} = \frac{1}{\# < 40}$. Between the two factions $\frac{1}{\# > 40}$, and $\frac{1}{\# < 40}$, $\frac{1}{\# < 40}$ is larger because its denominator $(\# < 40)$ is smaller than the denominator $(\# > 40)$ of $\frac{1}{\# > 40}$.

So $\frac{2}{81} < \frac{3}{111}$ or $\frac{83}{81} < \frac{114}{111}$.

6.8 Rationalization of factions

Q1.

Rationalizing denominators of the fractions by multiplying the fractions by $\frac{\sqrt{2}}{\sqrt{2}}$, we get $\frac{\sqrt{2}}{\sqrt{2}} \times \left(\frac{4}{3\sqrt{2}} - \frac{2}{\sqrt{2}}\right) = \frac{4.\sqrt{2}}{3.\sqrt{2}.\sqrt{2}} - \frac{2.\sqrt{2}}{\sqrt{2}.\sqrt{2}} = \frac{4\sqrt{2}}{3.2} - \frac{2\sqrt{2}}{2} = \frac{4\sqrt{2}}{6} - \frac{2\sqrt{2}}{2} =$ $\frac{\cancel{4}^{2}\sqrt{2}}{\cancel{6}_{3}} - \frac{\cancel{2}\sqrt{2}}{\cancel{2}} = \frac{2\sqrt{2}}{3} - \sqrt{2} = \frac{2\sqrt{2} - 3\sqrt{2}}{3} = \frac{-\sqrt{2}}{3}$.

=> $-\frac{\sqrt{2}}{3}$.

Q2.

$\frac{\sqrt{50}}{\sqrt{10}} \times \frac{\sqrt{18}}{\sqrt{12}} = \frac{\sqrt{5^2.2}}{\sqrt{5.2}} \times \frac{\sqrt{3^2.2}}{\sqrt{2^2.3}} = \frac{5\sqrt{2}}{\sqrt{5}\cancel{\sqrt{2}}} \times \frac{3\cancel{\sqrt{2}}}{2\sqrt{3}} = \frac{15\sqrt{2}}{2\sqrt{3}\sqrt{5}}$; to rationalise the denominator, we multiply the faction with $\frac{\sqrt{3}.\sqrt{5}}{\sqrt{3}.\sqrt{5}}$.

=> $\left(\frac{\sqrt{3}\sqrt{5}}{\sqrt{3}\sqrt{5}}\right) \times \left(\frac{15\sqrt{2}}{2\sqrt{3}\sqrt{5}}\right) = \frac{\cancel{15}\sqrt{2}\sqrt{15}}{2.\cancel{15}} = \frac{\sqrt{30}}{2}$.

6.9 Linear and Quadratic Equations

Q1.

$$x - 2y - 1 = 0 - - - - - - -(1)\ \&$$
$$-3y - x - 1 = 0 - - - - - - -(2)$$

By multiplying eqn (2) with '-1', we get $3y + x + 1 = 0$.

Let us calculate the value of x in terms of y from the first equation and substitute it in the second equation.

$x - 2y - 1 = 0$
=> $x = 1 + 2y$

By plugging in x in eqn. (2), we get,

$3y + (1 + 2y) + 1 = 0$

$3y + 1 + 2y + 1 = 0$

$=> 5y = -2$

$=> y = -\frac{2}{5}.$

By plugging in the value of $y = -\frac{2}{5}$ in the equation $x = 1+2y$, we get $x = 1+2\times-\frac{2}{5} = \frac{1}{5}$

So, $x = \frac{1}{5}$, and $y = -\frac{2}{5}$.

Q2.

$\frac{x}{2} - \frac{y}{3} + 4 = 0$ ——(1)

$x - \frac{y}{2} = -7$ ——(2)

There could be couple of approaches to solve these types of linear equations; we suggest that it is better to get rid of the denominator-constants from both the equations. The calls for taking LCM.

Taking LCM of the first eqn., and simplifying, we get,

$\frac{x}{2} - \frac{y}{3} + 4 = 0 => 3x - 2y + 24 = 0$ - - -(3)

Taking LCM of the second eqn., and simplifying, we get,

$x - \frac{y}{2} = -7 => 2x - y = -14$ - - -(4)

Now, we can calculate the value of y in terms of x from eqn. 4, so we get,

$2x - y = -14$ - - -(4)

$y = 2x + 14$ - - -(4)

By substituting the value of y in the first eqn, we get,

$3x - 2y + 24 = 0$ - - -(1)

$=> 3x - 2(2x + 14) + 24 = 0$

$=> 3x - 4x - 28 + 24 = 0$

$=> -x = 4 => x = -4.$

By plugging in the value of $x = -4$, in $y = 2x + 14$, we get,

$y = 2x + 14 \Rightarrow y = 2 \times -4 + 14 \Rightarrow y = -8 + 14 = 6 \Rightarrow y = 6$

So, $x = -4$, and $y = 6$.

Q3.

$2x^2 + 7x + 3 = 0$

Split the middle term $7x$ in two parts such that the product of the parts equals to $2x^2 \times 3 = 6x^2$.

$2x^2 + 7x + 3 = 0$

$\Rightarrow 2x^2 + 6x + x + 3 = 0$; splitting $7x$ into $6x$ and x works as $6x \times x = 6x^2$; though $7x$ can be split into $4x$ and $3x$, $4x \times 3x \neq 6x^2$.

Taking $2x$ common from first two terms, we get,

$2x(x + 3) + 1(x + 3) = 0$

Taking $(x + 3)$ common, we get,

$(x + 3)(2x + 1) = 0$

Product of two terms would be 0, if either of the term is 0. So, either $x + 3 = 0$ or $2x + 1 = 0$.

If $x + 3 = 0 \Rightarrow x = -3$ and if $2x + 1 = 0 \Rightarrow x = -\frac{1}{2}$.

So, $x = -3$ or $-\frac{1}{2}$.

Q4.

This is indeed a quadratic equation, though it does not look like one as we do not see a squared term; however the equation can be solved by applying the concept of quadratic equation.

Say $\sqrt{x} = y$, then the equation becomes $y^2 - 3y + 2 = 0$. It is a quadratic equation. Let's solve this.

$y^2 - 3y + 2 = 0$

$y^2 - 2y - y + 2 = 0$

$y(y-2)-1(y-2)=0$

$(y-2)(y-1)=0$

So, $y = 2$ or 1.

Or, $\sqrt{x} = 2$ or 1 or $x = 4$ or 1.

Q5.

By now you may be convinced that it is a quadratic equation. Let us make it look like one.

$x^{2/3} + x^{1/3} - 2 = 0$

$\Rightarrow (x^{1/3})^2 + x^{1/3} - 2 = 0$

Say $x^{1/3} = y$, then the equation becomes $y^2 + y - 2 = 0$.

Now solving $y^2 + y - 2 = 0$ should be a cake walk for you!

$y^2 + y - 2 = 0$

$\Rightarrow y^2 + 2y - y - 2 = 0$

$y^2 + 2y - y - 2 = 0$

$\Rightarrow y(y+2) - 1(y+2) = 0$

$\Rightarrow (y+2)(y-1) = 0.$

So, $y = -2$ or 1.

Or, $x^{1/3} = -2$ or 1

$x^{1/3} = -2$ or $1 \Rightarrow x = (-2)^3$ or 1^3

$x = -8$ or 1.

Let us check whether the values calculated are correct. Let us check for $x = -8$.

By plugging in the value of $x = -8$ in the equation $x^{2/3} + x^{1/3} - 2 = 0$, we get,

$(-8)^{2/3} + (-8)^{1/3} - 2 = 0$

$(-2^3)^{2/3} + (-2^3)^{1/3} - 2 = 0$

$(-2)^{3\times2/3} + (-2^{3\times1/3}) - 2 = 0$

$(-2)^2 + (-2^1) - 2 = 0$

$4 - 2 - 2 = 0 \Rightarrow 0 = 0$. The value calculated is correct.

Q6.

We know that for a quadratic equation $ax^2 + bx + c = 0$ will have equal roots if the discriminant $\Delta = b^2 - 4ac = 0$.

Here, $a = 1, b = 2(m - 4) = 2m - 8$, and $c = 2m$.

So, $\Delta = b^2 - 4ac = (2m - 8)^2 - 4.1.2m = 0$

$4m^2 - 32m + 64 - 8m = 0$

$4m^2 - 40m + 64 = 0$

$4(m^2 - 10m + 16) = 0$

$m^2 - 10m + 16 = 0$

$m^2 - 8m - 2m + 16 = 0$

$m(m - 8) - 2(m - 8) = 0$

$(m - 8)(m - 2) = 0$

or, $m = 8$ or 2.

Q7.

$(4x - 1)^2 = 81$.
Taking the root of both the sides, we get $\sqrt{(4x-1)^2} = \sqrt{81} \Rightarrow (4x - 1) = \pm 9$.

Remember that square root of any number always takes $\pm$ values.

Taking $+9$ first as the value, we get $(4x - 1) = 9 \Rightarrow x = 5/2$.

Now, Taking -9, we get $(4x - 1) = -9 \Rightarrow x = -2$.

So, = 5/2 or −2.

Q8.

$(a-3)(a+2)=(a+3)(a-2)$

=> $a^2+2a-3a-6=a^2-2a+3a-6$

=> $-a=a$

=> $2a=0$ => $a=0$.

So a will have only one unique value. Note that the answer is 'One' and not '0'. '0' is the value of a, whereas the question asks: how many unique values a can have?

Seeing a quadratic equation, one must not jump to a conclusion that the variable will necessary have two values!

Q9.

$a^3-a^2-a+1=0$

=> $a^2(a-1)-1(a-1)=0$

=> $(a-1)(a^2-1)=0$

=> $(a-1)(a-1)(a+1)=0$

=> $a=\pm 1$.

There are only two values of a : 1 or −1. The answer is: 'Two'. Again, seeing a cubic equation, one must not jump to a conclusion that the variable will necessarily have three values!

6.10 Absolute numbers

Q1.

$|10x|-20=0$

=> $|10x|=20$

=> $|x| = 20/10 = 2$

=> $|x| = 2$

=> $x = \pm 2$.

Q2.

$$\left[\frac{-|-3| \times -|3|}{(-|-3|)^2}\right]$$
$$= \left[\frac{-3 \times -3}{(-3)^2}\right]$$
$$= \frac{9}{9} = 1.$$

Q3.

$|2x - 5| = 8$ means that either $2x - 5 = +8$ or $2x - 5 = -8$

If $2x - 5 = +8$, then $2x = 8 + 5 = 13$ => $x = 13/2$.

And if $2x - 5 = -8$, then $2x = -8 + 5 = -3$ => $x = -3/2$.

So, $x = 13/2$ or $-3/2$.

6.11 Inequalities

Q1.

The set of values for the inequality: $1 \leq x < 3$ is infinite numbers.: {1, 1.1, 1.11, 1.111, 2, 2.345,...}; do not assume that x is an integer until stated. So x may have infinite number of values.

Q2.

Since LHS of the inequity $\frac{1}{p}$ is greater than 1, hence p must be positive.

We can manipulate the inequality as $\left(\frac{1}{p} \times p\right) > 1 \times p$

=> $1 > p$; since we know that p is a positive number, we can multiply both the sides of inequity with a positive number; had it been not known we cannot do it, so considering

p as positive, the range of p would be $0 < p < 1$.

Q3.

A. $ab > 0$

Inequality $ab > 0$ suggests that the product of a, and b is positive, it can be positive in two ways:

1. either both a, and b are positive

or

2. both a, and b are negative.

So, $a > 0$ and $b > 0$, **or** $a < 0$ and $b < 0$.

B. $ab < 0$

Inequality $ab < 0$ suggests that the product of a, and b is negative, it can be negative in two ways: either a is positive and b is negative or vice-versa.

So, $a > 0$ and $b < 0$, **or** $a < 0$ and $b > 0$.

C. $ab = 0$

Either $a = 0$ or/and $b = 0$.

D. $\frac{a}{b} > 0$

The results are same as (a). The inequality $\frac{a}{b} > 0$ suggests that the division of a, and b is positive, it can be positive in two ways:

1. either both a, and b are positive or

2. both a, and b are negative.

E. $\frac{a}{b} < 0$

The results are same as (b). The inequality $\frac{a}{b} < 0$ suggests suggests that the division of a, and b is negative, it can be negative in two ways: either a is positive and b is negative or vice-versa.

Q4.

$$\Rightarrow |2x - 1| < 5 \begin{cases} 2x - 1 < 5 \\ 2x - 1 > -5; \text{reverse the inequality sign to consider negative value.} \end{cases}$$

$$\Rightarrow |2x - 1| < 5 \begin{cases} 2x - 1 < 5 \Rightarrow 2x < 6 \Rightarrow x < 3. \\ 2x - 1 > -5 \Rightarrow 2x > -4 \Rightarrow x > -2. \end{cases}$$

So, the range for x is $-2 < x < 3$.

Q5.

$|x + 2| < 7$

$\Rightarrow |x + 2| < 7$ means that $-7 < x + 2 < 7$

$\Rightarrow -7 - 2 < x < 7 - 2$; subtracting 2 from each term

$\Rightarrow -9 < x < 5$

Q6.

$|1 - 2x| < 9$

$|1 - 2x| < 9$ means that $-9 < 1 - 2x < 9$

$-10 < -2x < 8$; Subtracting 1 from each term

$5 > x > -4$; Dividing each term by -2. The sign of inequality will change when we multiply or divide terms of inequalities by a negative number.

So, $-4 < x < 5$.

Q7.

$$\frac{5x - 1}{-6} \leq \frac{1 - 2x}{3}$$

$$\frac{5x-1}{-\cancel{6}^{\,2}} \le \frac{1-2x}{\cancel{3}}$$

$$\Rightarrow \frac{5x-1}{-2} \le 1-2x$$

$\left(\frac{5x-1}{-2}\right) \times -2 \ge -2 \times (1-2x)$; Sign of inequality will reverse as we multiplied it with a negative number ' – 2'.

$$\left(\frac{5x-1}{\cancel{-2}}\right) \times \cancel{-2} \ge -2 \times (1-2x);$$

$5x - 1 \ge -2 + 4x \Rightarrow 5x - 4x \ge -2 + 1 \Rightarrow x \ge -1.$

Alternatively

$\frac{5x-1}{-6} \le \frac{1-2x}{3}$ can be written as $-\frac{5x}{6} + \frac{1}{6} \le \frac{1}{3} - \frac{2x}{3}$

$$\Rightarrow -\frac{5x}{6} + \frac{2x}{3} \le \frac{1}{3} - \frac{1}{6}$$

$$\Rightarrow \frac{-5x+4x}{\cancel{6}} \le \frac{2-1}{\cancel{6}}$$

$$\Rightarrow -x \le 1$$

$$\Rightarrow x \ge -1.$$

Q8.

This is a 'Could be true' type of question. Even if the option is true for even one circumstance, it is the correct answer. 'Could be true' type of questions are different from 'Must be true' type of questions, in which the condition given must be true for all the circumstances.

I. $x^2 < 2x$: For $1 < x < 2$, the inequality $x^2 < 2x$ is true, else not. Say $x = 1/2$, then $(1/2)^2 < 2.(1/2) \Rightarrow 1/4 < 1$; so it can be true.

   Alternatively, in case of inequality, if x is a positive number, which is true in the given inequality ($1 < x < 3$), then we can cancel x from both the sides. So, $x^{\cancel{2}} < 2\cancel{x} \Rightarrow x < 2$. $x < 2$ is within the range of $1 < x < 3$, so option (I) is correct.

II. $x^2 = 2x \Rightarrow x = 2$ (cancelling x from both the sides); $x = 2$ is within the range of $1 < x < 3$, so option (II) is also correct. We must not even consider $x = 0$ as $x = 0$ is out of range of given constraints $1 < x < 3$.

III. $x^2 > 2x \Rightarrow x^{\not{2}} > 2\not{x} \Rightarrow x > 2$; Similarly, $x > 2$ is within the range of $1 < x < 3$, so option (III) is also correct.

The correct answer is option E.

Q9.

$ab^2c < 0$ suggests that LHS of the inequality: ab^2c is negative. Since b^2 is always positive irrespective of b being positive or negative, we cannot deduce whether b is positive or negative.

So, ab^2c is negative because ac is negative or $ac < 0$. This implies that either $a < 0$ and $c > 0$ or $a > 0$ and $c < 0$.

So, $ab^2c < 0$ only suggests that a, and c have opposite signs.

Q10.

First, you must match the direction of the inequalities. Currently, the directions of inequalities: $8 < x < 3$ and $4 < y < 6$ are opposite. $4 < y < 6$ can be written as $-4 > -y > -6$.

Adding the inequalities, we get...

$$
\begin{array}{l}
\quad 8 > \quad x > 3 \\
(+)\ -4 > -y > -6 \\
\hline
4 > x - y > -3 \\
\hline
\end{array}
$$

So the range of $(x - y)$ is $4 > x - y > -3$.

Q11.

$x^2 > -6x - 5$

$x^2 + 6x + 5 > 0 \Rightarrow x^2 + 5x + x + 5 > 0 \Rightarrow x(x+5) + 1(x+5) > 0 \Rightarrow (x+5)(x+1) > 0$.

$(x + 5)(x + 1)$ would be positive if either both $(x + 5)$ and $(x + 1)$ are negative or both are positive.

It is clear that $(x + 5)$ and $(x + 1)$ are positive if $x > -1$.

And, $(x + 5)$ and $(x + 1)$ are negative if $x < -5$.

So, $x > -1$ or $x < -5$.

Q12.

$x^2 > 6x - 5$

$x^2 - 6x + 5 > 0 \Rightarrow x^2 - 5x - x + 5 > 0 \Rightarrow x(x - 5) - 1(x - 5) > 0 \Rightarrow (x - 5)(x - 1) > 0.$

$(x - 5)(x - 1)$ would be positive if either both $(x - 5)$ and $(x - 1)$ are negative or both are positive.

It is clear that $(x - 5)$ and $(x - 1)$ are negative if $x < 1$.

And, $(x - 5)$ and $(x - 1)$ are positive if $x > 5$.

So, $x < 1$ or $x > 5$.

6.12 Percents

Q1.

Let us translate this mathematically: $24 = 8\%$ of x; say the number is x.

$\Rightarrow 24 = \frac{8}{100} \times x$

$\Rightarrow x = \frac{24 \times 100}{8} = 300.$

$\Rightarrow x = 300$

Q2.

Say, he examines x number of screws, so 0.03% of $x = 300$, or $x = 1,000,000$ screws.

Q3.

$x\%$ of $17 = 30\%$ of 289

$x\% \times 17 = 30\% \times 289$

$x\% \times 17 = 30\% \times 289^{17}$

$x = 510$

Q4.

Instead of calculating the discount and deducting it from \$800, it is better we calculate 60% of \$800: (100% - 40%) of the final price. So 60% of \$800 = $\frac{60}{100} \times 800 = \$480$.

Q5.

$\% \text{ loss} = \frac{\text{SP} - \text{CP}}{\text{CP}} \times 100\% = \frac{100 - 120}{120} \times 100\% = -16.67\%$; profit or loss are always based on Cost Price and not on Sales Price.

Q6.

The Marked Price would be 100 + 20 = \$120.

$\% \text{ Discount} = \frac{\text{Discount}}{\text{Marked Price}} \times 100\% = \frac{20}{120} \times 100\% = 16.67\%$.

Q7.

Say $Y = 100$, then $X = (100 + 110)\%$ of $100 = 210$. Note that Y is not more by 10%, it is more by 110%.

Given: $Z = X - 10\%$ of $X = 100\%$ of $X - 10\%$ of $X = 90\%$ of $210 = 189$.

Say X is $p\%$ of Z, then $p = \frac{X}{Z} \times 100\% \Rightarrow p = \frac{210}{189} \times 100\% = 111.11\%$.

Q8.

Say Number = N.

$N_1 = 1.20N$; N_1 is the number after the increase. Increasing an number by 20% can better be done by making it 1.20 times.

$N_2 = 0.80N_1 = 0.80 \times 1.20N = 0.96N = 96\%$ of N; N_2 is the number after the decrease. Decreasing an number by 20% can better be done by making it 0.80 times. Remember that the decrease will effect on N_1 and not on N.

The resulting number ($0.96N$) is not the same as the original number (N), it decreased by 4%.

What if the number were decreased first, and then increased later?

Well the result would have been the same: A 4% decrease!

Q9.

Instead of solving it with a traditional approach, you can apply See-Saw technique discussed earlier in the book.

Step 1: Convert % into fraction. $25\% = \frac{1}{4}$;

Step 2: Add '1' to the denominator of the fraction. $\frac{1}{4+1} = \frac{1}{5}$;

Step 3: Convert the fraction into %. $\frac{1}{5} = 20\%$.

The shopkeeper should offer a discount of 20% so that he does not run into losses.

6.13 Ratio & Proportion

Q1.

We know that "Product of middle terms = Product of extreme terms".

$5 \times x = 0.25 \times 4 \Rightarrow 5x = 1/4 \times 4 \Rightarrow 5x = 1 \Rightarrow x = 1/5$.

Q2.

Let the third number be 100.

Then, first number = 120% of 100 = 120.

Second number = 150% of 100 = 150.

Ratio of first two numbers = 120 : 150 = 12 : 15 = 4 : 5.

Q3.

The given ratio can be simplified as $\frac{1}{2} : \frac{2}{3} : \frac{3}{4} = \frac{6}{12} : \frac{8}{12} : \frac{9}{12} = 6 : 8 : 9$; ignoring the LCM = 12.

Third part = $529 \times \frac{9}{6+8+9} = 529 \times \frac{9}{23} = 207$.

Q4.

Given is: $\frac{3}{5}$ of A's amount = $\frac{2}{5}$ of B's amount

Or, $\left(\frac{3}{\not{5}}\right) A = \left(\frac{2}{\not{5}}\right) B \Rightarrow 3A = 2B$.

=> A : B = 2 : 3 on ratio scale.

<u>Actual scale</u> <u>Ratio scale</u>
A : B : A + B :: A : B : A + B
? : ? : 240 :: 2 : 3 : 5

So, $A = \frac{240}{5} \times 2 = 96$ & $B = \frac{240}{5} \times 3 = 144$.

Q5.

Let the numbers be $3x$ and $5x$, where x is a common factor between the numbers.

Then, $\frac{3x-9}{5x-9} = \frac{12}{23}$

$\Rightarrow 23(3x - 9) = 12(5x - 9) \Rightarrow 9x = 99 \Rightarrow x = 11$.

The larger number = $5x = (5 \times 11) = 55$.

Q6.

Simply plug in the ratio values of A, B, and C.

$= \frac{A}{B} : \frac{B}{C} : \frac{C}{A} = \frac{2}{3} : \frac{3}{4} : \frac{2}{1}$

$= \frac{8}{12} : \frac{9}{12} : \frac{24}{12}$; taking the LCM of 3, 4, and 1 and then ignoring it.

= 8 : 9 : 24.

Q7.

Quantity of milk = $60 \times \frac{2}{3}$ liters = 40 liters.

Quantity of water in the mixture = (60 – 40) liters = 20 liters.

New ratio of milk : water = 1 : 2 (Reversed)

Let the quantity of water to be added further be x liters.

Then, milk : water = $40 : (20 + x)$

$1 : 2 = 40 : (20 + x)$

Now, $\frac{40}{20 + x} = \frac{1}{2} \Rightarrow 20 + x = 80 \Rightarrow x = 60$.

Quantity of water to be added = 60 liters.

Alternate approach

Since the revised ratio of milk and water is 1 : 2, and milk is not added, it remains 40 liters as before

$\Rightarrow 40 = \left[\frac{1}{2+1}\right]$ of new mixture

So, the new mixture = $40 \times \frac{3}{1} = 120$ liters, or quantity of water = 120 – 40 = 80 liters. It means that quantity of water added = 80 – 20 = 60 liters.

Q8.

Given: $A : B :: 2 : 3; B : C :: 2 : 3$; and $A + B + C = 196$.

B has two values on ratio scale: '3' and '2', we must have a unique value for B for the ratio $A : B : C$.

Let us take B as '3' on ratio scale, then $B : C :: 2 : 3 = 3 : \left(\frac{3}{2} \times 3\right) = 3 : \frac{9}{2}$. Note that C changes to $\frac{9}{2}$.

So, $A : B : C = 2 : 3 : \frac{9}{2} \Rightarrow A : B : C = 4 : 6 : 9$.

Third number = $190 \times \left(\frac{9}{4 + 6 + 9}\right) = 190 \times \left(\frac{9}{19}\right) = 90$.

Q9.

Say the current salaries of A, B, and C are $2k, 3k$, and $5k$, respectively, whether k is a common factor among the salaries.

A's new salary $= \frac{115}{100} \times 2k = \frac{23}{10}k;$

B's new salary $= \frac{110}{100} \times 3k = \frac{33}{10}k;$

C's new salary $= \frac{120}{100} \times 5k = 6k$

New ratio $= \frac{23k}{10} : \frac{33k}{10} : 6k = 23 : 33 : 60.$

Q10.

Let the numbers be $2x, 3x$ and $5x$.

Then, $(2x)^2 + (3x)^2 + (5x)^2 = 1862$

=> $4x^2 + 9x^2 + 25x^2 = 1862$

=> $38x^2 = 1368$

=> $x^2 = 36$ => $x = 6$.

=> First number $= 2x = 2.6 = 12$.

6.14 Simple Interest

Q1.

$$\text{SI} = \frac{P \times r \times t}{100}$$

$$\Rightarrow \text{SI} = \frac{8,000 \times 50/3 \times 27/12}{100}$$

$$\Rightarrow \text{SI} = \frac{\cancel{8,000}^{\cancel{80}^{20}} \times 50 \times \cancel{27}^{\cancel{9}^{3}}}{\cancel{3} \times \cancel{12}^{3} \times \cancel{100}} = \$3,000.$$

=> SI = \$3,000.

Q2.

$$SI = \frac{P \times r \times t}{100} \Rightarrow P = \frac{SI \times 100}{r \times t}$$

By applying above formula we can easily solve this question, as we are already having the simple interest.

$$\Rightarrow P = \frac{1,230 \times 100}{6 \times 5} \Rightarrow P = \$4,100.$$

Q3.

Let the sum 'P' be \$100, so it becomes 'A = \$400' in 15 years or SI= 400 – 100 = \$300.

Now, $r = \frac{SI \times 100}{P \times t}$

$$\Rightarrow r = \frac{300 \times 100}{100 \times 15} = 20\%.$$

$\Rightarrow r = 20\%$.

Q4.

Gain = $\left(\frac{P \times r_1 \times t}{100}\right) - \left(\frac{P \times r_2 \times t}{100}\right)$, where $P = 5,000$, $t = 2$, $r_1 = 25/4\%$, and $r_2 = 4\%$.

$$\Rightarrow \text{Gain} = \left(\frac{5,000 \times 25/4 \times 2}{100}\right) - \left(\frac{5,000 \times 4 \times 2}{100}\right) = 625 - 400 = \$225.$$

$\Rightarrow$ Gain = \$225.

Q5.

From the question, we can conclude that $2,200 = \frac{P_1 \times r \times t_1}{100} + \frac{P_2 \times r \times t_2}{100}$; rate is same for both B and C.

$$\Rightarrow 2,200 = \frac{5,000 \times r \times 2}{100} + \frac{3,000 \times r \times 4}{100}$$

$$\Rightarrow 2,200 = 100r + 120r \Rightarrow 2,200 = 220r \Rightarrow r = 10\%.$$

6.15 Compound Interest

Q1.

We know that $A = P\left(1+\frac{r}{100}\right)^n$; where A = Amount, P = Principal or Sum.

$A = 7,500\left(1+\frac{4}{100}\right)^2$

$= 7,500\left(\frac{26}{25}\right)^2$

$= 7,500 \times \frac{26}{25} \times \frac{26}{25}$

$= \$8,112.$

So, compound interest $= A - P = 8,112 - 7,500 = \$612.$

Q2.

$CI = A - P$

$CI = P\left(1+\frac{r}{100}\right)^n - P$

$=> 4,000\left(1+\frac{10}{100}\right)^2 - 4,000$

$=> 4,000\left(\frac{11}{10} \times \frac{11}{10} - 1\right)$

$=> 4,000(1.21 - 1)$

$=> \$840.$

So, SI = 1/2 of CI $= \frac{840}{2} = 420$

Or, P $= \frac{SI \times 100}{r \times t}$

$= \frac{420 \times 100}{3 \times 8}$

$P = \$1,750.$

Q3.

$$SI = \frac{P \times 4 \times 2}{100} = \frac{2}{25}P$$

$$CI = A - P$$

$$CI = P\left(1 + \frac{4}{100}\right)^2 - P$$

$$= \frac{676P}{625} - P$$

$$= \frac{51P}{625}$$

As, C.I. – S.I. = 1

$$=> \frac{51P}{625} - \frac{2P}{25} = 1$$

$$=> \frac{51P - 50P}{625} = 1$$

$=> P = \$625.$

Alternate approach

It is to be noted that for the first year there is no difference between SI & CI. CI start increasing compared to SI from the second year onwards.

In the question, the difference between CI and SI amounting to \$1 for two years is due to the INTEREST on first year SI for one year.

$$1 = \frac{SI_{I\text{ yr.}} \times 4\% \times 1^{st}\text{ yr.}}{100} => SI_{I\text{ yr.}} = \$25$$

$$\text{Again, } SI_{I\text{ yr.}} = 25 = \frac{P \times 4\% \times 1}{100} => P = \$625$$

Q4.

Let the principal be \$1 for 1 year when compounded half yearly; so $n = 1 \times 2 = 2$ periods; Rate $r = 20/2 = 10\%$ half yrly.

$$A = P\left(1 + \frac{r/2}{100}\right)^{2n}$$

Amount $A = 1\left(1 + \frac{10}{100}\right)^2 = (1.1)^2 = 1.21.$

So, Effective rate annual rate of interest = $(1.21 - 1) \times 100\% = 21\%$ p.a.

Q5.

It is to be noted that the difference between simple interest and compound interest is: In case of SI, the interest for any year is calculated on the fixed sum invested initially, whereas in case of CI, the interest for each subsequent year is calculated on the amount of the previous year.

In the question, 840 is the amount for the 4th year and 800 is the amount for the 3rd year, which means that $840 - 800 = 40$ is the interest for the 4th year (one year) on the amount \$800.

So, $r = \frac{SI \times 100}{P \times t}$

$=> r = \frac{40 \times 100}{800 \times 1}$

$=> r = 5\%.$

Alternate approach

$$\frac{\text{Amount}_{n^{th}\ yr.}}{\text{Amount}_{(n-1)^{th}\ yr.}} = 1 + \frac{\text{Rate}}{100}$$

So, $\frac{840}{800} = 1.05$; or Rate = 5%.

6.16 Functions

Q1.

Since $f(x) = 2x - 7$, we have $f(2x - 6) = 2.(2x - 6) - 7 = 4x - 12 - 7 = 4x - 19$; replacing '$x$' with $2x - 6$.

Now, $3f(x) - 3 = f(2x - 6) \Rightarrow 3.(2x - 7) - 3 = 4x - 19$; substituting the values of $f(x) = 2x - 7$ and $f(2x - 6)$.

We get, $6x - 21 - 3 = 4x - 19 \Rightarrow x = 5/2$.

Q2.

$[25] = -|\sqrt{25}| = -5$; as 25 is an odd number.

And, $[10] = \dfrac{10}{5} = 2$; as 10 is an even number.

So, $[25] \times [10] = -5 \times 2 = -10$.

Q3.

Set of integers less than -2.2 are {-3, -4, -5,....}; among them, the greatest is -3, so, $[-2.2] = -3$.

Similarly, set of integers less than 2.6 are {2, 1, 0, -1,....}; among them, the greatest is 2, so, $[2.6] = 2$.

Similarly, set of integers less than 6.2 are {6, 5, 4,....}; among them the greatest is 6, so, $[6.2] = 6$.

So, $[-2.2] + [2.6] + [6.2] = -3 + 2 + 6 = 5$.

Q4.

$a\#c = \dfrac{(a + c)}{(a^2 + c^2)} = 0 \Rightarrow a + c = 0 \times (a^2 + c^2) \Rightarrow a + c = 0 \Rightarrow c = -a$.

Q5.

Say $f(x) = y$, then we can say that $f(y) = -1$ and deduce $f(y) = \dfrac{2y}{2y - 1} = -1 \Rightarrow -2y + 1 = 2y \Rightarrow y = 1/4$.

Again, $y = f(x) = \frac{1}{4} => \frac{2x}{2x-1} => 2x - 1 = 8x => x = -1/6.$

Q6.

$$f(g(x)) = \frac{\cancel{2}.\left(\frac{2x-1}{\cancel{2}x}\right)}{\cancel{2}.\left(\frac{2x-1}{\cancel{2}x}\right)-1} = \frac{\left(\frac{2x-1}{x}\right)}{\left(\frac{2x-1}{x}\right)-1} = \frac{\left(\frac{2x-1}{\cancel{x}}\right)}{\left(\frac{2x-1-x}{\cancel{x}}\right)} = \frac{2x-1}{x-1}.$$

Now, $f(g(x)) = 3 => \frac{2x-1}{x-1} = 3 => 2x - 1 = 3x - 3 => x = 2.$

Q7.

$f(p^2 - 10p + 25)$ can be simplified as $f((p - 5)^2)$.

So, $f(p^2 - 10p + 25) = f((p-5)^2) = 2.(p-5)^2 + \sqrt{(p-5)^2}$

Taking $\sqrt{(p-5)^2} = p - 5$, we get $f(p^2 - 10p + 25) = 2p^2 - 20p + 50 + p - 5 = 2p^2 - 19p + 45.$

Taking $\sqrt{(p-5)^2} = 5 - p$, we get $f(p^2 - 10p + 25) = 2p^2 - 20p + 50 - p + 5 = 2p^2 - 21p + 55.$

$=> f(p^2 - 10p + 25) = 2p^2 - 19p + 45$ or $2p^2 - 21p + 55.$

Note: $\sqrt{(p-5)^2}$ can be $p - 5$ or $5 - p$ depending on whether $p > 5$ or $5 > p$; since, by definition, the 'Root' symbol only takes the positive root.

Had we known that $p > 5$, then $\sqrt{(p-5)^2}$ would have been only $(p - 5)$.

Chapter 7

Speak to Us

Have a Question?

Email your questions to *info@manhattanreview.com.* We will be happy to answer you. Your questions can be related to a concept, an application of a concept, an explanation of a question, a suggestion for an alternate approach, or anything else you wish to ask regarding the SAT.

Please do mention the page number when quoting from the book.

Best of luck!

Professor Dr. Joern Meissner
& The Manhattan Review Team

Made in United States
North Haven, CT
05 November 2026

21763963R00102